COMITE EURO-INTERNATIONAL DU BETON

SEISMIC DESIGN

OF REINFORCED CONCRETE STRUCTURES FOR CONTROLLED INELASTIC RESPONSE

DESIGN CONCEPTS

Thomas Telford

Published by Thomas Telford Ltd, 1 Heron Quay, London E14 4JD, UK, for the Comité Euro-International du Béton, Case Postale 88, CH-1015 Lausanne, Switzerland.

First published 1997 as CEB Bulletin d'Information No. 236 *Seismic design of reinforced concrete structures for controlled inelastic response: recent advances in design concepts and codes.*
Thomas Telford edition published 1998
URL:http://www.t-telford.co.uk

Distributors for Thomas Telford books are
USA: ASCE, 1801 Alexander Bell Drive, Reston, VA 20191-4400
Japan: Maruzen Co. Ltd, Book Department, 3–10 Nihonbashi 2-chome, Chuo-ku, Tokyo 103
Australia: DA Books and Journals, 648 Whitehorse Road, Mitcham 3132, Victoria

SEISMIC DESIGN—240
This CEB Bulletin 240 is the 1998 revised edition of the former Bulletin 236 which was first published in 1977 under the same title. It continues the series of hard-cover CEB technical monographs.

The book is presented to subscribing, associate and sponsoring *fib* members as part of their 1998 subscriptions.

A catalogue record for this book is available from the British Library

ISBN: 0 7277 2641 2

Typeset by MHL Typesetting, Coventry.
Printed in Great Britain by The Cromwell Press, Trowbridge, Wilts.

Acknowledgements

This Design Guide has been written by the members of CEB Task Group III/2: *Seismic design of reinforced concrete structures* and was approved by CEB Commission III *Design*.

Convenor:	Paolo E. Pinto	Italy
Members:	Daniel P. Abrams (since 1996)	USA
	Michele Calvi	Italy
	Amr Elnashai	United Kingdom
	Michael N. Fardis	Greece
	Andreas Kappos	United Kingdom
	Camillo Nuti	Italy
	Voula Pantazopolou (since 1996)	Canada
	Manuel Pipa	Portugal
	Nigel Priestley	USA
	Günter Waas	Germany
	Franz A. Zahn	Switzerland
Corresponding members:	Filip C. Filippou	USA
	Shunsuke Otani	Japan

Chapters drafted by:

1. Objective and scope — Paolo E. Pinto
2. Code design procedures — Amr Elnashai, Shunsuke Otani, Paolo E. Pinto, Nigel Priestley, Franz A. Zahn
3. Reliability based system analysis — Paolo E. Pinto, Felice Colangelo, Renato Giannini
4. Measures of seismic performance — Andreas Kappos
5. Selected case studies — Amr Elnashai, Michael N. Fardis, Andreas Kappos
6. Assessment of existing buildings — Michele Calvi, Nigel Priestley

Contents

1. Objective and scope

1.1. Philosophy of seismic design for reinforced concrete structures

For a structure to remain elastic under its design seismic action, typically associated with a 10% exceedance probability in 50 years, it has to be designed for lateral forces with magnitude in the order of 50% or more of its weight. Although technically feasible, designing a structure to respond elastically to its design seismic action is economically prohibitive. It is also completely unnecessary, as the earthquake is a dynamic action, representing for a structure a certain total energy input and a demand to tolerate a certain level of displacement and deformation, but not a demand to withstand specific forces. Therefore, seismic design codes allow the development of significant inelastic response under the design seismic action, provided that the magnitude of inelastic deformations does not endanger the integrity of the individual members and of the structure as a whole.

Pending development of simple and reliable procedures for seismic design on the basis of the displacement seismic demands, and as structural design has been traditionally force-based, structures are still designed for earthquake resistance by proportioning their members for internal forces computed from a linear elastic analysis of the response to specified lateral forces.

These design forces are obtained from a design acceleration spectrum, usually derived by dividing the elastic spectral accelerations by the 'behaviour factor' q (of European seismic design codes) or the 'force reduction factor' R (of North American codes).

Individual members of the structure are proportioned for resistance to the internal forces derived from the (inelastic) design spectrum, and, in addition, detailed to develop the inelastic deformations associated with the value of the q or R factor. This is achieved if the structure is designed to develop a global displacement ductility factor μ_δ at the top (defined as the peak response displacement at the point of application of the resultant lateral force, divided by the corresponding displacement at yielding) at least equal to that associated with the value of q or R, on the basis of which its members are proportioned.

For flexible structures, with natural periods in the constant velocity or constant displacement regions of the spectrum, the 'equal displacement rule' applies in good approximation, giving $\mu_\delta = q$ for the demand value of the global displacement ductility factor. For stiffer structures, with periods up to the constant acceleration region, the requirement for equal deformation energy in the cycle of the largest deformation excursion, elastic or inelastic, gives $\mu_\delta = (q^2 + 1)/2$.

The desired global displacement ductility demand is distributed as uniformly as possible to all members and regions of the structure capable of developing inelastic deformations in a ductile manner. Given such a distribution, the global displacement ductility demand is translated into local rotation and ductility demands of those members and regions entrusted to develop inelastic action. Finally, these latter regions are detailed so that they can reliably and safely sustain the corresponding local ductility demands.

1.2. Capacity design and ductility classes

In multi-storey buildings, spreading the inelastic deformation demands uniformly throughout the structure means mobilizing all storeys into the inelastic action. In RC structures this can be achieved only if vertical members (columns and walls) remain essentially elastic in all storeys, with the exception of the base of the bottom storey.

In that region significant rotations are allowed to develop, either through stable and controlled plastic hinging of the vertical members, or through soil deformations and rocking of foundation elements on the ground. Under these conditions, kinematics dictate that inelastic lateral displacements will be associated with the development of plastic hinges at the end regions of nearly all the beams of the building, a situation which corresponds to the most uniform spreading of inelastic deformation demands possible in the structures.

Modern seismic design standards strive to distribute ductility demands to all beams of the structure and to avoid formation of a soft storey, by forcing vertical elements to remain elastic with the exception of their base region. This is achieved by overdesigning (relative to demands of the elastic analysis for the design seismic action) the columns in bending and the entire height of shear walls in comparison with their base region. Overdesign of columns is typically achieved through application of the corresponding *capacity design* rule. According to this rule, beam sections at column faces are proportioned strictly on the basis of the moments obtained from the linear-elastic analysis of the structure for the combination of the design seismic action with the simultaneously acting quasi-permanent gravity loads, while internal force demands in the end sections of the columns framing into a joint are derived, not from the linear-elastic analysis as above, but from the actual flexural capacities of the beams framing into the same joint, in such a way that simultaneous plastic hinging at the base of the column above and at the top of the column below the joint is precluded.

In spite of the apparent simplicity of the concept, a full implementation of the capacity design approach, so as to ensure the formation of a specific dissipative inelastic mechanism, leads to rather complex procedures. In fact, not only has one to increase the spreading of inelastic demands to, and to enhance the ductility capacity of, the flexural elements, but at the same time premature failure of other mechanisms or elements, such as shear, beam–column joints, connections, foundations, etc., which would reduce the efficiency of the main mechanism must be avoided.

The complexity of the procedures obviously increases with the amount of ductility one wishes to exploit.

The incentive towards an increase of ductility supply is the fact that design forces vary almost inversely proportional to it.

Limiting the magnitude of lateral forces that a structure can sustain has a number of advantages. Foundation structures are lighter, the forces transmitted to the soil are lower, which reduces the likelihood of permanent deformations, the maximum response accelerations of the structure are also lower, affording better protection to any equipment mounted on the structure which may be sensitive to acceleration. Besides, within ample limits, the displacements of a yielding structure are nearly independent of the amount of inelastic action. Ductility is, finally, the most effective defence against unanticipated, unfavourable characteristics of the ground shaking.

In several cases, however, there are good reasons for the alternative choice, namely less ductility and more strength. These are the cases of structures that, because of other permanent or variable loads which they have to resist, and because of their moderate height and the level of hazard to which they are exposed, naturally possess a significant fraction of the strength required to resist the design seismic action. In these cases, since detailing for strength is easier and more economical than detailing for ductility, it is quite justified to renounce some ductility at the expense of some extra strength.

Most modern seismic codes provide for more than one combination of strength and ductility: this is the case of the codes of New Zealand, the USA, Japan and Europe. For example, the European seismic code: Eurocode 8 (EC8) allows for three alternative ductility classes. Each class has its own design forces, capacity design rules and factors, and rules for detailing. With some exaggeration, one might say that the collective provisions given for a class constitute a separate code in themselves.

1.3. Scope of the Design Guide

The two preceeding introductory paragraphs are essentially making the point that codified seismic design has developed, over a short period of time, to a relatively high level of sophistication.

The design forces depend on the ductility supply, local and global. Local ductility is a matter of type and amount of detailing. Global ductility is a matter of structural configuration and, given this latter, of the relative proportioning of the strength of the elements, so that inelastic demand goes to the largest number of those elements which can easily be made to be dissipative.

The whole is a complicated mechanism, whose functioning depends on the effectiveness of a large set of provisions and factors. Add to this the freedom that it is appropriate to extend to the user of selecting different combinations of strength and ductility, with the overall safety level remaining unchanged.

The rapidity of the development which has occurred in seismic codes might leave one with the suspicion that not everything has been properly considered, and that some aspects of the process may have gone out of control. This is certainly not the case, since codes are written by large bodies of experts and, nowadays, there is always international cross-checking.

The need is felt, however, and not only by the users of codes, to review the theoretical framework underlying the process of modern seismic design, if for no other reason than to make it more transparent, and explain its logic to people outside the small circle of experts.

This Design Guide aims to satisfy this perceived need. The way in which it attempts to do this is basically a pragmatic one. It begins by outlining briefly the approaches taken by four major regional codes, followed by a commentary on similarities and points of difference among them.

Next comes an essay on what could be today a feasible higher-level approach, in which uncertainties and variabilities are modelled in a probabilistic context and the results are expressed in an explicit reliability format. The purpose of the essay is a demonstrative one, to indicate that the problem can be theoretically formulated in a comprehensive way, and results of practical value can be obtained, although with a number of limitations.

The bulk of the Design Guide is then devoted to reasoned applications, consisting of designs of structures performed according to different criteria, and subsequently checked by means of non-linear analyses. The comments provided on the different assumptions made in the examples, the explanations given during the design and verification processes, the discussions on a comparative basis of the results obtained, should all enable the interested reader to understand the relative weight of the numerous parameters involved and the interrelationships between them, as well as the motivations behind the choices adopted by the codes.

The Guide concludes with a second essay, containing a double element of novelty: one of substance; the second of method. The subject is the extension of the system viewpoint to the assessment of existing structures, a crucial topic in the future of earthquake engineering. The method is the so-called displacement-based approach, whereby one starts from an estimated ultimate deformation mechanism to arrive at the seismic intensity capable of producing it.

This last chapter has been included to serve as a bridge towards future research activities already being planned.

2. Code design procedures

2.1. Introduction

Engineers involved in code-making for earthquake resistant design have shown themselves to be capable of considerable ingenuity in devising effective procedures for satisfying their objectives on the basis of little more than common sense. And the objectives in question were far from simple ones: to control the behaviour of a structure acted upon by a ground motion of very variable and unpredictable characteristics, intense enough to bring the structure close to, but not beyond, the exhaustion of its inelastic deformation capacity.

This chapter contains an overview of the procedures developed for the above purpose in four regions of the world, which are distant from each other both physically and in their cultures. The intent is not to describe these procedures in full, the reader is presumed to be familiar with at least one of them, but rather to reduce them to their essential features, in order to see as clearly as possible (some of them are not exactly straightforward) not just the differences between them, but the reasons behind these differences as well.

It can be said at this point that similarities dominate, and that wherever the rules differ, this is clearly because different objectives are pursued. This check has merits of its own, as proof of the existence of a single logic behind non-identical forms.

But the main purpose of this chapter is actually to prepare the reader for the following ones.

Starting from a *de facto* situation, i.e. the existence of effective procedures for controlling the dynamic inelastic response of structures, a theoretical substantiation is first sought (Chapter 3) by having recourse to stochastic random vibration methods, which were not available when the procedures were initially devised.

Value and limitations of the more fundamental approach are discussed with reference to the pragmatic solutions obtained via qualitative reasoning and traditional, unsophisticated analyses.

A deeper insight into the complex interrelationships existing among the capacity design parameters is provided in the following parts of the Design Guide. Chapter 5, for example, demonstrates the results of parametric analyses on real case structures having an unprecedented width of scope, one that only the present power of computing tools can permit. Again, without some previous training on the prescriptions contained in the codes, the mental gymnastics required to grasp all the implications emerging from these analyses would have been excessive.

The four national or regional codes dealing with capacity design will be presented following the historical order in which they have been implemented.

2.2. The New Zealand approach to capacity design for reinforced concrete structures

2.2.1. *Introduction*

The New Zealand capacity design principles for ductile moment-resisting RC frames aim primarily at

- establishing a strong column — weak beam structure, i.e. eliminating the possibility of a column sway mechanism (soft storey) even during the most severe seismic motions
- avoiding shear failures in columns and beams.

In order to avoid the formation of column plastic hinges (except at the base of the column and at roof level), inelastic dynamic effects, causing the bending moment diagrams in the columns to differ substantially from those derived from an elastic analysis based primarily on first mode response, must be taken into account.

Consequently, the capacity design procedures are mainly concerned with deriving column design actions consistent with large inelastic deformations that cause plastic hinges in the beams to develop overstrength moments, and considering inelastic dynamic effects. The beams are designed for shear forces consistent with the beam hinges at both ends developing overstrength moments.

For ductile structural walls and coupled walls, similar principles are involved with a requirement for a dependable inelastic mechanism based on wall base flexural hinges and coupling-beam hinges, together with a capacity design approach to avoid shear failures.

Generally, for both walls and frames, the principles can be summarized in a single general capacity design equation

$$\phi_S S_N \geq \omega_s \phi_0 S_E \tag{2.1}$$

where S_E is the required value of action S based on specified lateral loads, ϕ_0 is the overstrength factor associated with the design flexural hinge mechanism, ω_s is a dynamic amplification factor for action S, S_N is the nominal strength of action S and ϕ_S is the strength reduction factor for action S. In determining required flexural strength of plastic hinges, $\phi_o = \omega_s = 1{\cdot}0$.

In seismic design to the ultimate limit state (ULS), there is only one load combination

$$U = 1{\cdot}0G + 1{\cdot}0\Psi Q + 1{\cdot}0E \tag{2.2}$$

where G = dead load, Q = live load, Ψ is the reduction factor for ULS live load, normally taken as 0·4, and E is the earthquake design load. When using the equivalent static force method of design, the loading code specifies the following equation for the base shear

$$V = C_h(T_1, \mu)S_p RZW \tag{2.3}$$

In equation (2.3) C_h (T_1, μ) is the spectral coefficient dependent on the structural displacement ductility factor μ and on the first mode period of vibration T_1, calculated by the Rayleigh method. Three different site conditions are considered, with inelastic spectra plotted for values of $1 \leq \mu \leq 10$, based on equal energy considerations for short periods, and equal displacements for $T_1 \geq 0{\cdot}7$ s. The spectra are considered to be based on 5% damped elastic response ($\mu = 1$) corresponding to a 450-year return period. S_p is a structural performance factor, generally 0·67, R is a risk factor, Z is a zone intensity factor ($0{\cdot}6 \leq Z \leq 1{\cdot}2$), and W is the seismic weight. Modal analysis and time-history analysis techniques are allowed as an alternative to the equivalent static force method.

2.2.2. *Capacity design approach for frames*

The aim is to ensure, by suitable overstrength and dynamic amplification factors, that a beam sway mechanism forms, with plastic hinges forming in beams at carefully identified locations (generally but not exclusively the column faces, at column base, and possibly at the top of the upper storey columns). As with the EC8 approach, the intent is to spread plasticity to the largest possible number of hinges, and to minimize plastic rotations. The following approach is used to ensure this desirable behaviour:

2.2.2.1. Beams

2.2.2.1.1. Flexural strength

Design actions. Required dependable moments are those resulting from the static or dynamic lateral analysis, using the load combination of equation (2.2). Redistribution of bending moments reducing peak moments by up to 30% and increasing others is permitted, provided storey shear capacity is not reduced. Design moments are found by dividing the required moment by a strength reduction factor of $\phi_f = 0{\cdot}85$. Design capacity is based on characteristic material strengths.

2.2.2.1.2. Ductile behaviour. Prescriptive detailing requirements, involving spacing, amount and anchorage details of transverse reinforcement, and reinforcement ratios for longitudinal reinforcement are defined to ensure that a dependable structural displacement ductility capacity of $\mu \geq 6$ is assured. For frames designed for limited ductility (essentially $\mu \leq 3$), less stringent requirements related to confinement and antibuckling stirrups apply.

2.2.2.1.3. Shear strength

Design actions. Capacity design procedures implied by equation (2.1) are implemented by determining the overstrength factor ϕ_0 associated with development of plastic hinges. This is based on multiplying the design flexural strength, based on reinforcement provided, including tributary slab reinforcement, by a steel overstrength factor $\gamma = 1{\cdot}25$. Required beam shear strength is then found from equilibrium considerations related to overstrength moments at the plastic hinges, and distributed gravity loads on the beam. No strength reduction factor is applied to design shear strength, because of perceived conservatism in determining the possible magnitude of the shear action. In plastic hinge regions, strength of concrete shear resisting mechanisms is commonly ignored. Upper levels of design shear stress in ductile beams are set lower than for non-seismic conditions.

2.2.2.2 Columns. Capacity design procedures, in accordance with equation (2.1), are required for fully ductile frames. This involves determination of the period-dependent dynamic amplification factor ω as well as the overstrength factor ϕ_0. For limited ductile frames, simplified procedures are applied where the dynamic amplification factor is a constant for all frames.

2.2.2.2.1. Bending moment. Nodal overstrength factors ϕ_{0j} are determined separately for each beam–column joint by extrapolating the beam overstrength moments at plastic hinge locations to the joint centroids, and comparing the sum ΣM_{bo} of these from both beams (if an interior joint) framing into the joint with the sum ΣM_{bE} of seismic beam moments determined at the joint centroid under the specified lateral forces.

Thus

$$\phi_{0j} = \Sigma M_{bo} / \Sigma M_{bE} \tag{2.4}$$

With a perfect match of required and design beam flexural strength (with no redistribution), typically $\phi_{0j} = 1{\cdot}47$, but redistribution and lack of perfect strength match causes considerable variation in ϕ_{0j}. The dynamic amplification factor for columns depends on whether the column is part of a one-way or two-way structural frame system

$$\text{one-way frames:}\, 1{\cdot}3 \leq \omega = 0{\cdot}6T_1 + 0{\cdot}85 \leq 1{\cdot}8 \tag{2.5a}$$

$$\text{two-way frames:}\, 1{\cdot}5 \leq \omega = 0{\cdot}5T_1 + 1{\cdot}1 \leq 1{\cdot}9 \tag{2.5b}$$

The value of ω from equation (2.5) is reduced near the ground floor and top floor, where higher mode effects are less apparent, and $\omega = 1{\cdot}0$ applies at foundation level, since this is a designated plastic hinge location.

Column required strength at beam faces is found by application of equation (2.1), at the joint centroid using the above values for ϕ_{0j} and ω, and reducing the moment to account for probable column shear strength, taken as 0·6 V_{col}, where V_{col} is the overstrength (design) column shear strength, discussed below. Hence, the required column moment capacity is

$$M_{col} = \phi_{0j}\omega M_E - (V_{col})(0{\cdot}3h_b) \qquad (2.6)$$

where M_E is the column moment under specified lateral loads, and h_b is the beam depth.

Again the design strength is matched to the required strength, that is, no strength reduction factor is employed. It should be noted that the higher ω factor for two-way frames allows for the column to be designed uniaxially for flexure. Thus, biaxial flexural design is not required.

2.2.2.2.2. Axial force. Column axial force resulting from seismic action is calculated on the assumption of beam hinges forming at flexural overstrength in all beams at levels higher than that under consideration, and finding the algebraic sum of consequent beam seismic shear forces V_{EO} framing into the column. This sum is reduced dependent on the number of storeys above the level considered, the fundamental period T_1 and whether the column is part of a one-way or two-way structural system. Hence

$$N_{col,E} = R_v \Sigma V_{EO} \qquad (2.7)$$

where the reduction factor varies from 1·0 to 0·55 as the number of levels and ω increase.

2.2.2.2.3. Shear. Because column moment capacities are increased substantially from the elastic values M_E corresponding to the specified lateral forces, it is extremely unlikely that column plastic hinges can form at the top and bottom of a column between adjacent storeys, except, possibly, at ground floor level. Hence, although a dynamic amplification factor is appropriate for shear, it is less so than for column flexure. For all levels except ground-floor columns

$$\text{one-way frames: } V_{col} = 1{\cdot}3\phi_{0j}V_E \qquad (2.8a)$$

$$\text{two-way frames: } V_{col} = 1{\cdot}6\phi_{0j}V_E \qquad (2.8b)$$

The higher factor of $\omega = 1{\cdot}6$ for two-way frames allows the columns to be designed without consideration of biaxial interaction effects. The value of ϕ_{0j} is taken to be the average at the joints at top and bottom of the column under consideration. Column shear force for ground-floor columns is based on the assumption of plastic hinges at the top and bottom of the column. This does not, however, presuppose the development of a soft-storey mechanism, but is in recognition of the fact that beam growth under inelastic action at the first floor level may cause localized column hinging below the beams.

2.2.2.3. Beams–column joints. Maximum shear stress in beam–column joints is calculated based on the assumption of beam flexural reinforcement stress equal to 1·25 f_y, and is limited to 0·2 f'_c to avoid the possibility of joint diagonal compression failure. A portion of the shear force is assumed to be carried by concrete shear-resisting mechanism, with the amount varying between one-way and two-way frames. The remainder of the shear force must be carried by properly designed transverse and vertical joint reinforcement. Typically, the vertical reinforcement is provided by column longitudinal reinforcement which is under-utilized for flexure.

2.2.2.4. Foundations. Normally, foundations are required to remain elastic, and must therefore be designed for forces resulting from flexural overstrength of column base hinges. However, the possibility of limited ductile foundation structures, particularly foundation beams, based on a displacement ductility factor of $\mu = 3$ is recognized, in which case the foundation may be designed for actions resulting from the specified lateral forces, provided that appropriate detailing is employed.

2.2.3. Capacity design for wall structures

The aim is to ensure, by suitable overstrength and dynamic amplification factors, that a desirable inelastic deformation mechanism based on flexural ductility is ensured. Where the lateral force-resisting mechanism consists of linked vertical cantilever walls, plastic hinges must only develop at the wall base. For coupled structural walls, additional inelastic rotation is concentrated in the coupling beams. Since these are expected to be subjected to high ductility demand combined with moderate to high shear force, special detailing requirements, typically involving diagonal reinforcement in the coupling beam, are imposed.

2.2.3.1. Walls

2.2.3.1.1. Flexural strength. Required base shear strength is calculated using a maximum structural displacement ductility factor of $\mu = 5$ for multiple cantilevered walls and $\mu = 6$ for coupled walls. These values are reduced for walls with aspect ratios less than 3. Higher mode effects at upper levels of the wall are accommodated by assuming a linear envelope of required moment strength decreasing from the base moment value to zero at the wall top. Thus, dynamic amplification factors are not directly applied to wall moments in the form of equation (2.1). Tension shift of an amount equal to the wall length is assumed in determining locations for termination of longitudinal reinforcement.

2.2.3.1.2. Ductility capacity. Ductility capacity is assured by two main requirements. First, the wall thickness must be sufficient so that transverse buckling does not occur. The propensity for buckling is primarily related to the phase of compressing bars to remove previously developed inelastic tensile strain, and hence the requirements for wall width are a complex combination of wall aspect ratio, ductility level, and amount of reinforcement in the wall end. Secondly, when the longitudinal reinforcement ratio in wall ends exceeds about 0·5%, closely spaced transverse reinforcement, similar in amount and detailing to that required for columns, is required.

The coupling beams of ductile coupled walls must be reinforced in such a way that the entire earthquake-induced shear and flexure is resisted by diagonal reinforcement, unless the shear stress is less than

$$v_n = 0{\cdot}1 \frac{L_n}{h} \sqrt{f_c'} \text{MPa} \tag{2.9}$$

where L_n and h are the clear span and depth of the coupling beam. Special transverse reinforcement is required to restrain the reinforcement against buckling.

2.2.3.1.3. Shear strength. The design shear forces at all levels of structural walls are obtained from the basic capacity design equation

$$V = \omega_v \phi_0 V_E < \mu V \quad (2.10)$$

where the dynamic amplification factor for shear ω_v is related to the number of storeys N, by

$$N \leq 6 \quad \omega_v = 0{\cdot}9 + N/10 \quad (2.11a)$$

$$N > 6 \quad \omega_v = 1{\cdot}3 + N/30 \leq 1{\cdot}8 \quad (2.11b)$$

The flexural overstrength factor ϕ_o is the ratio of base overstrength moment capacity, with reinforcement stress = 1·25 f_y, to the moment resulting from specified lateral forces. With a minimum value for ϕ_0 of about 1·4, equation (2.11) results in design forces no less than 1·4 to 2·5 (depending on N) times the shear force corresponding to specified lateral forces. With the high capacity protection provided by this approach, no shear strength reduction factor is applied to the nominal shear strength. Reduced concrete contribution to shear strength applies in the hinge region.

Since shear and flexural strength in coupling beams are carried by the same reinforcement, no capacity protection is applied to shear force in these members.

2.2.4. Diaphragms

The important action of floor slabs in transmitting seismic forces to lateral force-resisting elements by diaphragm action is emphasized, particularly for wall structures, where diaphragm spans can be large. Capacity design procedures are required to determine the diaphragm forces to be carried elastically by the floor slabs. These forces can be particularly high for dual systems (mixed frame/wall structures). Particular attention is drawn to force transfer between slabs and walls.

The requirements of NZ S 3101 relating to diaphragms are descriptive, relating to basic capacity principles, rather than prescriptive, involving design equations.

2.3. Seismic design approach in the USA

2.3.1. Existing seismic codes and guidelines

There are a number of codes dealing with seismic resistance in the USA alongside documents that may be viewed as providing source material for code-drafting. In this section non-exhaustive notes are given on most of the existing codes and guidelines for the purpose of setting the scene for a more detailed assessment of the capacity design-related content of two leading documents.

There are four codes dealing with seismic provisions for buildings in the USA (NIST, 1992). These are the Council for American Building Officials (CABO) code for dwellings, the Building Officials and Code Administrators (BOCA) National Building Code, the Southern Building Code Congress International (SBCCI) Standard Building Code and the International Conference of Building Officials (ICBO) Uniform Building Code. Whereas the first three codes have adopted, to a greater or lesser degree, the provisions of the National Earthquake Hazard Reduction Program (NEHRP) guidelines, the latter code is traditionally linked to the guidelines by the Structural Engineers Association of California (SEAOC). Therefore, reviewing the Uniform Building Code and the NEHRP guidelines would provide a reasonable representation of existing seismic design practice in the USA. Emphasis is placed on the NEHRP guidelines since it is anticipated that these will form the basis for a national code.

The National Earthquake Hazard Reduction Program (NEHRP) is managed by the USA Federal Emergency Management Agency (FEMA). The provisions developed within the program are the responsibility of the Building Seismic Safety Council (BSSC) with FEMA funding and control. The most recent seismic design guidelines published as NEHRP documents are the 1994 edition (FEMA, 1995), which is the third such provision. Whereas NEHRP guidelines are not codes, and hence are not legally binding, they are increasingly seen as the leading source documents for code development. It may be the case that the target 'international code' under consideration in the USA will be largely based on the NEHRP provisions operative at the time.

A brief review of the 1994 Seismic Regulations for New Buildings follows with regard to capacity design and controlled inelastic response of RC structures only. Since the provisions provide clauses additional to, or in place of, guidelines given by ACI 318–89 (1992 edition), the latter code is also reviewed briefly. Recommended response modification factors are also reviewed since they have a direct effect on local ductility demand (in beams), hence their values influence the column protection factors. The loading and classification parts are presented to serve the above purpose and are not intended to be comprehensive. It should also be noted that seismic provisions of the Uniform Building Code (UBC, 1994) are similar to ACI 318–89 and often to NEHRP. Differences from the NEHRP approach are identified in the following section.

2.3.2. Seismic action

2.3.2.1. Design ground motion. Design ground accelerations are specified in the form of maps giving the effective peak acceleration A_a and the effective velocity-related acceleration A_v. These are presumed to provide uniform hazard for a 50-year design life with 10% probability of being exceeded. A_a and A_v range between 0·05 and 0·4 for continental USA. The Uniform Building Code identifies only one acceleration coefficient, similar to A_v in application, ranging between 0·075 and 0·4.

2.3.2.2. Soil profile type. The site coefficients F_a and F_v are evaluated according to one of six soil profiles, A to F; A being rock and F being potentially liquefiable or high plasticity soil for which special studies are needed. Three measures of soil characteristics are used; shear wave velocity, standard penetration and undrained shear strength. F_a and F_v values are $0{\cdot}8 \div 2{\cdot}5$ and $0{\cdot}8 \div 3{\cdot}5$ respectively. Since F_v is operative for longer period structures, it follows that seismic forces for longer period structures will be higher than those evaluated according to earlier regulations. The UBC identifies four soil conditions. Multiplication factors to the basic coefficient for S1 to S4 soils range from 1·0 to 2·0 but are capped at lower periods by a constant peak response level.

2.3.2.3. Seismic forces. The seismic design force (for equivalent static load analysis) is given by

$$V = C_s W$$

where

$$C_s = 1{\cdot}2C_v/RT^{2/3}$$

$$C_v = F_v A_v$$

Alternatively, C_s need not exceed the value

$$C_s = 2{\cdot}5C_a/R$$

where C_s is the seismic response coefficient, R is the response modification factor (behaviour factor), T is the fundamental period of the structure, F_a and F_v are site response coefficients and C_a and C_v are seismic coefficients (products of F and A). The two-parameter approach above was developed to account for the high amplification on rock sites, the very high amplification on soft sites for low levels of shaking and the reduction in amplification on soft sites as the severity of shaking increases. It also caters for the general perception that long period structures are more vulnerable to instability.

The UBC-94 approach is similar, when an 'equivalent lateral force' approach is used. The seismic design force is given by

$$V = (ZIC/R_w)W$$

where

$$C = 1{\cdot}255/T^{2/3} \leq 2{\cdot}75$$

In the above, $0{\cdot}075 \leq Z \leq 0{\cdot}4$ is the zone factor, and I is an importance factor; generally $I = 1$ but for essential facilities $I = 1{\cdot}25$. The period T is calculated by a simple equation of the form

$$T = 0{\cdot}0731(h_n)^{3/4} \quad \text{for concrete frame structures}$$

$$T = 0{\cdot}0488(h_n)^{3/4} \quad \text{for concrete wall structures}$$

where h_n is the height in metres. Alternatively, the period may be calculated using Rayleigh's method, but the results are not permitted to deviate from the above equation by more than 30%.

A modal dynamic analysis may be also used to determine V, incorporating sufficient modes to capture 90% of the total mass. Results are assessed using a different spectrum than that corresponding to the above equation for C and often the results are very different. In such cases, the calculated base shear is factored up to correspond to the basic approach (not the Rayleigh equation) multiplied by 0·8. Thus, the main effect is to modify the shape of the lateral force vector.

2.3.2.4. Response modification and displacement amplification factors R and C_d. The elastic forces are scaled down by the response modification factor to arrive at design forces, in recognition of ductile response and damping. The R factor (analogous to q in EC8) is considered to be an integral part of the capacity design framework, since its value implies a certain level of beam and column base ductility demand, and hence should affect the necessary level of column overstrength. It is clearly stated in the Commentary that it is empirical and intended to represent only ductile response and damping. As such, the R values range from 1·25 for non-ductile unreinforced masonry and non-ductile steel moment frames to 8·0 for special moment frames and steel and RC. It is noted that these values are lower than the UBC R_w values, which go up to 12, in recognition of the working stress basis of R_w. For ultimate strength design, UBC specifies a load factor of 1·4. Thus the maximum R factor would be 12/1·4 = 8·4. On the other hand, the minimum force reduction factor has decreased from 4 in UBC to 1·25 in NEHRP, an observation which increases confidence in the values given since it is difficult to justify a reduction of 4 for a non-redundant and brittle structure.

Alongside the definition of R values for various structural systems and materials, structural system and height limitations are given for each seismic performance category (defined below). Also, C_d factors used to amplify computed displacements for purposes of drift checks and in *P–d* calculations are given. The C_d factors are either less than or equal to the *R* factors in NEHRP and even less in UBC. It is likely that the C_d factors will underestimate the building displacement (ATC-19, 1995).

2.3.3. *Load combinations*

The load combinations are defined in the Commentary to the NEHRP provisions. These are given as

$1{\cdot}2D + 1{\cdot}0E + 0{\cdot}5L + 0{\cdot}2S$ when seismic and gravity forces are additive
$0{\cdot}9D + 1{\cdot}0E$ when they are counteractive

where *D* is for dead load, *E* is the seismic load, *L* is the live load and *S* is the snow load. The seismic load is given by

$E = Q_E \pm 0{\cdot}5C_aD$ + for additive and − for counteractive

where *E* is the effect of horizontal and vertical earthquake-induced forces and Q_E is the effect of horizontal seismic forces. For columns supporting discontinuous lateral force-resisting elements, the load combination is made more onerous by introducing a capacity design related parameter (2*R*/5), leading to

$$E = (2R/5)Q_E + 0{\cdot}5C_aD$$

The above expression takes into account that higher ductility demands are imposed on beams in high *R* designed structures, hence the column overstrength should be increased accordingly. No importance factors are considered in seismic load combinations. This was agreed after careful consideration of field evidence and the observation that increased levels of design forces do not necessarily provide higher seismic protection. In this context, more stringent detailing requirements are more effective than increased design loads.

As previously mentioned, UBC-94 load factors multiply the earthquake load by 1·4, effectively reducing the impact of the large R_w specified. The basic UBC-94 equations are

$$U = 1{\cdot}4\ (D + L + E)$$

and

$$U = 0{\cdot}9D \pm 1{\cdot}4E$$

where *L, D* and *E* are as defined above.

2.3.4. *Seismic performance categories*

Central to design and detailing recommendations is the selection of the appropriate seismic performance category. This is based on the seismic hazard exposure group and the velocity-related ground acceleration A_v as given in Table 2.1. Seismic hazard exposure group III is for essential facilities, II for substantial public importance and I for other structures.

Table 2.1 Definition of seismic performance categories

	Seismic hazard exposure group		
	I	II	III
$A_V < 0{\cdot}05$	A	A	A
$0{\cdot}05 \leq A_v < 0{\cdot}10$	B	B	B
$0{\cdot}10 \leq A_v < 0{\cdot}15$	C	C	C
$0{\cdot}15 \leq A_v < 0{\cdot}20$	C	D	D
$A_v \geq 0{\cdot}20$	D	D	E

Seismic performance category E is the most onerous in terms of regularity and detailing, while performance category A is for structures with little or no special seismic performance measures. All local and global supply and demand requirements are linked to the seismic performance category into which the structure falls (which is analogous to the EC8 ductility classes). The UBC does not have seismic performance categories which are, however, partly covered by the importance factor I.

2.3.5. *Seismic design requirements for reinforced concrete structures*

Reference is made to ACI 318–89 (1992 edition) with replacements and modifications given below. Only those clauses relevant to controlled inelastic deformations are cited.

2.3.5.1. Classification of moment frames. The main thrust of the provisions is the detailing of moment frames according to the selected seismic performance category (A, B, C, D or E). The three classes are

(*a*) ordinary moment frames — frames designed and detailed to ACI 318–89 with no seismic provisions
(*b*) intermediate moment frames — frames designed and detailed to ACI 318–89 section for structures in areas of moderate seismic hazard, in addition to all requirements of ordinary moment frames
(*c*) special moment frames — frames designed and detailed to the full special seismic provisions of ACI 318–89, Chapter 21 in addition to the requirements of (*a*) above.

Unlike Eurocode 8, no explicit capacity design requirements are imposed on frames according to the target seismic performance level. Capacity design is implicit in prescribing that certain performance categories dictate the use of certain types of frame. Category A frames may be ordinary moment frames. Category B frames, which form part of the seismic load resistance system, should be intermediate frames. Category C frames may be either intermediate or special moment frames, while for categories D and E only special moment frames are allowed.

Since ACI 318–89 does not deal with seismic design of precast concrete structures, the provisions give guidance on design of structures with strong and ductile connections. These have a clear capacity design objective. They are, however, not reproduced here since equivalent clauses do not exist in other codes reviewed within this chapter.

In addition to the provisions of ACI 318–89, brief comments are given for the design of coupling beams in coupled wall structures with bi-diagonal reinforcement. To ensure that the flexural capacity of coupling beams is adequately assessed for shear reinforcement, the provisions require that the contribution of inclined bars to the flexural strength is included in design shear force calculations.

2.3.6. *ACI 318–89 seismic design requirements*

Since the NEHRP provisions are intended for use alongside ACI 318–89 (Chapter 21 of the 1992 edition), clauses from the latter relevant to controlled inelastic response are reviewed below.

2.3.6.1. General. The code does not explicitly state that a desired failure mode is aimed at, neither does it give an indication of where inelastic action is undesirable. It is stated that the code is intended for use alongside an acceptable code for loading which defines levels of seismic risk. Reference is made to possible compatibility with SEAOC, ATC (No. 510) and UBC. A word of caution is given in the Commentary regarding possible conflict between the various loading codes.

All structures in areas of low seismic risk need not be designed specifically for earthquake resistance according to Chapter 21 of the code. Moreover, walls, diaphragms and trusses in moderate seismic risk areas are also exempt.

There is a fundamental contradiction between the ACI 318-89 and NEHRP provisions in so far as the trade-off between strength and ductility is concerned. Whereas the latter state that increased design loads do not necessarily imply commensurate increase in seismic safety, the former state the opposite. ACI 318–89 goes further in stating in the Commentary that ductile response is the aim for structures in high seismic risk areas while strength design is the objective of the ACI clauses for areas of moderate seismic risk.

2.3.6.2. Capacity design requirements. There are five aspects in the code that may be classified strictly under capacity design. These are as follows.

- The probable flexural strength, used to evaluate design shear forces in the cases below, is evaluated using a steel yield stress of 1·25 times the specified yield stress and no strength reduction factors.
- For structures in high seismic risk areas, the design shear force is evaluated not from applied design loads but from the probable flexural strength (defined above) of the member under consideration. In areas of moderate risk, the design shear force is the largest of (*a*) that corresponding to nominal flexural strength and (*b*) from analysis under factored loads. Walls and diaphragms are exempt even in high seismic risk areas.
- When the axial load on frame members is less than $f'_c \, A_g/20$, the contribution of axial load to concrete shear resisting mechanism is ignored.
- The actual yield strength of tensile reinforcement should not be more than 18 000 psi higher than the specified value. A further 3000 psi violation is permitted upon retesting.
- The sum of column design flexural strength at a joint should be equal to or greater than 1·2 times the sum of moments corresponding to flexural strength of the beams. This may be violated if any positive effect of the columns is neglected, all negative effects are catered for and additional confinement reinforcement is placed up the full column height.

Other guidelines in the code are intended to develop inelastic action in members and ensure a level of ductility, but are not strictly capacity design related. These are reviewed below, noting that compatibility with NEHRP requires that ACI clauses for structures in areas of moderate seismic risk are interpreted as pertaining to intermediate moment-resisting frames.

2.3.6.3. Ductility of beams

- The clear span should be $\geq$ four times the member depth (for flexural response to prevail).
- Hoops should be provided for a distance of twice the member depth from the face of the joint at each end (with the first hoop not more than 2" from the joint face) and on either side of yielding sections for the same distance.

- Maximum hoop spacing not exceeding (*a*) *d*/4, (*b*) 8 times the smallest bar diameter of longitudinal reinforcement, (*c*) 24 times the hoop diameter or (*d*) 12".
- Maximum hoop spacing throughout not more than half the member depth.
- If the shear force due to earthquake loading is $\geq$ 50% of the total shear, the concrete contribution to the shear resistance should be neglected.

2.3.6.4. Ductility of columns

- Volumetric ratio of spiral hoops not less than 0·12 times the ratio of the concrete compressive strength to the hoop yield stress. Expressions for the minimum cross-sectional area of rectangular hoops and their acceptable configurations are also given.
- Hoop spacing for structures in high risk areas should not exceed (*a*) one-quarter of the minimum member dimension or (*b*) 4". For moderate risk areas, the spacing should not exceed that for beams given above. This should be provided for a length of not less than (*i*) member depth at location of inelasticity, (*ii*) one-sixth the member length or (*iii*) 18" in all cases, with the first hoop at half the minimum spacing. For areas of moderate seismic risk, hoops at spacing not exceeding twice the above spacing should be provided throughout.
- For columns supporting discontinuous stiff members, stirrups should be provided for the whole length of the member below the connection with the discontinuous member if the axial load in the member exceeds 10% of the product of the gross concrete area and its unfactored compressive strength.

2.3.6.5. Design and detailing of joints

- Actions on joints should be determined from member forces assuming a yield stress of 1·25 times the specified yield stress.
- Joints not confined by members, according to the dimension ratio member to joint, should be provided with hoops.
- Joint width $\geq$ 20 times the largest beam bar diameter crossing the joint.

2.3.7. Commentary

The NEHRP provisions, while developed as an independent document, seek to progress the use of the existing materials codes (e.g. AISC and ACI). As such, they are restricted in developing or adopting new concepts. Therefore, it seems that an explicit and complete capacity design framework is still under consideration. There are aspects of the provisions that are of interest, in comparison to other codes, such as the decision not to increase the seismic force as a function of building importance, but to rely more on detailing, which is contrary to the philosophy of ACI 318–89. Also, the brief rules given for precast concrete structures and coupling beams in coupled wall structures do exhibit an explicit recognition of capacity design.

Notwithstanding the above, the provisions indirectly include capacity design of moment frames insofar as they make reference to ACI 318–89. In the latter code, there are several clauses that are capacity design inspired, as presented above. It is, however, concluded that NEHRP, used alongside ACI 318–89, provides some ingredients of a capacity design approach, but this is not as complete as other modern codes, such as the New Zealand code reviewed in section 2.2.

2.4. The Eurocode 8 approach to capacity design for reinforced concrete structures

2.4.1. Introduction

Eurocode 8 (1994) allows design to be carried out using three different balances between the strength and the ductility of a structure: the higher the ductility, the lower the design forces, and hence the strength of the structure. The three levels, labelled as DCH, DCM and DCL, are intended to be equivalent in terms of seismic protection with respect to the ultimate limit state. Owing to the fact that exploitation of ductility is different for the three DCs, however, capacity design provisions also need not be the same for the three DCs, their severity decreasing with the lowering of ductility demand. Actually, the capacity design provisions for DCL are rather minimal.

The flexibility of EC8 with respect to admissible structural ductilities, a necessity for making it usable in countries with large differences in seismicity, has required a greater effort, for the absolute and relative calibration of the capacity design procedures and factors, than is required for other codes. Calibration studies have been and are currently being conducted by a number of investigators: examples of these studies are presented in Chapter 5. The preliminary indications coming from these studies are that EC8 designed structures of different ductility levels are actually equivalent from the point of view of safety; it also appears (a fact which was not necessarily to be expected) that the trade-off between strength and ductility is feasible at constant cost, at least regarding the quantities of materials to be employed (the difference in workmanship required, for example for DCM and DCH, may have weight).

2.4.2. Design seismic action as function of ductility class

The seismic action in EC8 is defined by means of an elastic response spectrum, whose shape is regulated by six parameters to fit regional features and whose scaling factor is the peak ground acceleration (PGA) appropriate for the particular zone. For structures of ordinary importance, a return period of 475 years is indicated as a basis for the determination of the PGA value.

The design spectrum is obtained from the elastic one, essentially by dividing the ordinates by a constant factor q, called the behaviour factor, and related to ductility class and structural typology. Frames and coupled walls structures of DCH have a q-value of 5, which is reduced to 4 for uncoupled walls structures. For the two lower DCs the q-values are multiplied by a factor of 0·75 and 0·5, respectively, yielding values of q equal to 3·75 and 2·5.

The design seismic load combination is as follows

$$G_k + \gamma_I E + \Sigma\psi_{2i} Q_{ik}$$

where G_k and Q_k are the characteristic values of permanent and variable loads, respectively, ψ_{2i} are reduced combination coefficients (of the order of 0·3 for ordinary variable loads, like occupancy loads, etc.), and γ_I is an importance factor which increases the seismic effects with respect to the reference ones, so as to make them correspond to selected (higher than 475 years) return periods.

2.4.3. Capacity design approach for frames

The capacity design approach of EC8 for frame structures aims at a well defined global mechanism, i.e. the so-called beams-sway mechanism, in which all beams at all storeys form plastic hinges, while all columns remain elastic for their entire height, with the exception of their base section at the ground storey.

In this mechanism, the global inelastic drift (plastic displacement at the top divided by the height of the frame) equals the plastic rotations at the end regions of all the beams, which corresponds to the most uniform possible spreading of inelasticity throughout the structure.

To achieve this objective, the following provisions for the proportioning of the various elements are given.

2.4.3.1. Beams

2.4.3.1.1. Bending

Action effects. The design values of the bending moments are those obtained from the analysis of the structure for the seismic load combination. This applies to all three DCs.

Ductile behaviour. Adequate flexural ductility is ensured by the detailing rules adopted, in particular those related to the minimum/maximum percentages of reinforcement, to the minimum amounts of reinforcement required all over the length of the beam, and to the transverse confining reinforcement at the 'critical' regions, whose length is to be taken as equal to $2h_w$, $1{\cdot}5h_w$ and $1{\cdot}0h_w$ for DCs H, M and L, respectively (h_w is the height of the beams). The spacing of the hoops and the tension reinforcement ratio in the critical regions are also different for the three DCs.

2.4.3.1.2. Shear

Action effects. Capacity design approach is only required for DCH; in the remaining two cases of DCM and DCL the design shear forces are taken directly from the analysis of the structure for the seismic load combination.

For DCH the design shear forces are obtained from the equilibrium condition of the beam subjected to the appropriate values of permanent and variable loads and to end resisting moments corresponding to the actual reinforcement provided, further multiplied by a factor $\gamma_{Rd} = 1{\cdot}25$. This factor is intended both to compensate for the safety factor $\gamma_m = 1{\cdot}15$ applied to the yield strength of steel, and to account for the variability of the actual strength, as well as for a degree of strain hardening.

At each end of the beam, two values of the design shear forces are to be calculated, corresponding to the possible double combination of positive and negative resisting moments at the two ends.

Shear capacity. It is well known that the shear capacity of beams and beam–columns is a (decreasing) function of the required flexural ductility: the procedures for its evaluation must therefore be differentiated according to the DC chosen.

One can note that in this case capacity design needs to consider not just the action effects to be used for dimensioning of the non-ductile elements, but the corresponding resistances as well. Eurocode 8 requirements are given below.

Beams of DCH

(*a*) The contribution of concrete to shear resistance is disregarded.

(*b*) If the maximum and minimum values of the shear force at one section are of the same sign, or if reversal occurs, but one has $|V_{max}| \leq 3(2 + \zeta)\tau_{Rd}bd$, where $\zeta = V_{min}/V_{max}$ and τ_{Rd} depends on the concrete class according to Eurocode 2 (*Reinforced concrete*): the shear reinforcement is evaluated on the basis of the truss model as given in EC2.

(*c*) If reversal occurs and $|V_{max}|$ exceeds the value above, bi-diagonal reinforcement at $\pm$ 45° is required, the relative amount of which depends on the value of $|V_{min}|$.

Beams of DCM
(*a*) The contribution of concrete to shear resistance is 40% of that provided in EC2.
(*b*) The same rules as for DCH apply, with the limit for $|V_{max}|$ increased to 4/3 of the previous value.

Beams of DCL. The shear resistance is evaluated according to EC2.

2.4.3.2. Columns

2.4.3.2.1. Bending and axial force. The determination of the design action effects on columns makes use of capacity design procedures only for DCH and DCM structures, the difference between the two cases being restricted to the numerical values of the 'overstrength factor' γ_{Rd}.

The objective of avoiding column hinging is deemed to be adequately achieved by satisfying the condition that at each beam–column joint the sum of the bending resistances of the column end sections is larger than the corresponding one of the beams.

Denoting by ΣM_{Sb} and ΣM_{Rb} the sum of the bending moments at the end sections of the beams framing into a generic joint, as obtained from the analysis of the structure under the design load combination (ΣM_{Sb}), and evaluated at the same sections on the basis of the actual reinforcement provided (ΣM_{Rb}), and by ΣM_{Sc} and ΣM_{Rc} the corresponding quantities for the end sections of the columns framing into the same joint, the condition previously expressed can be formulated as

$$\Sigma M_{Rc} \geq \gamma_{Rd}\Sigma M_{Rb} \not> q\Sigma M_{Sc} \qquad (2.12)$$

which can alternatively be formulated as

$$\Sigma M_{Rc} \geq \frac{\gamma_{Rd}}{\delta}\Sigma M_{Sc} \not> q\Sigma M_{Sc} \qquad (2.13)$$

with

$$\delta = \frac{\Sigma M_{Sc}}{\Sigma M_{Rb}} = \frac{\Sigma M_{Sb}}{\Sigma M_{Rb}}$$

with γ_{Rd} fulfilling the same functions already explained for the case of the beams, and whose values are γ_{Rd} = 1·35 and 1·20 for DCH and DCL, respectively.

Equations (2.12) and (2.13) are to be evaluated for both signs of the earthquake action in a given direction and, if the column belongs to two orthogonal frames (two-way column), for both orthogonal directions. In each direction, the most unfavourable situation (i.e. the largest value of the 'global overstrength factor' γ_{Rd}/δ) is assumed.

Biaxial bending must be taken into account in the dimensioning/verification of the columns: one option allowed by EC8 is to take the full action effect due to one of the two horizontal components and to combine it with 30% of the action effect due to the second horizontal component (and then exchange the order): in each case the overstrength factor to be used is the one related to the dominant action in the combination.

Concerning the evaluation of the axial force, EC8 relies on the results of the structural analysis, by simply stating that the variation of the axial force due to the earthquake action has to be taken into account for the purpose of 'determining the most unfavourable combination of bending and axial force'.

The capacity design rule given by equation (2.12) or (2.13) becomes unnecessarily severe when bending moments in beams due to gravitational loads are relatively important compared with those due to lateral actions.

Under these circumstances, the positive bending moments at the beams ends due to the seismic actions may not lead to the inversion of the sign of the negative moments due to the vertical loads or, even if they do, the resultant positive moment may be less than the available positive resisting moment (according to EC8 the bottom reinforcement must be in all cases no less than half of the top reinforcement required in beams to resist the negative bending moment).

In these cases it is unduly conservative to ask that the column end sections at a joint take up the sum of the negative resisting bending moment of the beam at one side *and* the positive one of the beam at the other side.

To correct this situation, Eurocode 8 introduces an alternative format to that of equation (2.12) or (2.13), which is more versatile and can be applied in all cases.

The formulation is

$$\Sigma M_{\mathrm{Rc}} \geq (1 + \gamma_{\mathrm{Rd}} - \delta)\delta \Sigma M_{\mathrm{Rb}} = (1 + \gamma_{\mathrm{Rd}} - \delta)\Sigma M_{\mathrm{Sc}} \not> q \Sigma M_{\mathrm{Sc}} \quad (2.14)$$

When full inversion takes place, δ is close to unity, and the two formats of equation (2.12) and (2.14) are equivalent. When, on the contrary, ΣM_{Sb} is significantly less than ΣM_{Rb}, equation (2.14) provides a reduction of the global overstrength factor, whose lower bound is $(1 + \gamma_{\mathrm{Rd}})$.

Columns designed according to the capacity design procedure (which is nominally intended to avoid their hinging) must still be endowed with an adequate reserve of ductility. An expression is given in EC8 for the amount of transverse reinforcement necessary to achieve prescribed values of 'curvature ductility'.

These values are 13, 9 and 5 for columns of DCH, M and L, respectively. Also, the length within which the confining reinforcement is needed varies with the ductility class.

Development of plastic hinges at the base section of the ground-storey columns is explicitly accepted in the global mechanism of the frame. To avoid hinging in the columns at the base sections occurring before that of the beams at the various storeys, EC8 prescribes application at the bottom section of an 'overstrength factor' equal to that calculated for the top section.

2.4.3.3. Shear. Capacity design shear forces are evaluated in EC8 assuming that plastic hinges form at both ends, i.e. by means of the expression

$$V = \gamma_{\mathrm{Rd}} \frac{M_{\mathrm{Rc1}} + M_{\mathrm{Rc2}}}{h} \quad (2.15)$$

where $M_{\mathrm{Rc1,2}}$ are the flexural capacities of the end sections as detailed (the earthquake action has to be considered with both signs), h is the clear height of the column and γ_{Rd} = 1·35 and 1·2 for DCH and M, respectively.

No capacity design is required for columns of DCL. The design flexural capacities $M_{\mathrm{Rc}i}$ in equation (2.15) should be the maximum possible, considering the variation of the columns axial force due to the seismic action. Since, however, shear capacity as given in EC2 is an increasing function of axial force, the most critical situation of shear demand and shear capacity must in general be found by more than one trial.

2.4.3.4. Beam–column joints. The shear forces acting around the joints are evaluated by considering capacity design conditions for the concurring beam end sections, using γ_{Rd} factors of 1·25 and 1·15 for DCH and DCM, respectively. Explicit checking of DCL joints is not required.

2.4.3.5. Foundations. Design action effects for foundation structures and foundation soil are obtained on the basis of capacity design resistances of the superstructure (but they need not be larger than the elastically calculated ($q = 1$) seismic effects).

2.4.4. Capacity design for wall structures

The design value of the bending moment at the wall base is the one obtained from the analysis of the structure under the respective seismic load combination, for all DCs.

To account for a possible overstrength at the base section, which would increase the wall moments with respect to those from the analysis, as well as to account for inelastic higher mode effects for slender walls (i.e. those with $h_w/l_w > 2$), the design bending moment diagram is a linear envelope of the calculated one, displaced vertically by a distance equal to the 'critical region' of the wall: h_{cr}. The value of h_{cr} is the maximum between the width of the wall and 1/6 of its height.

This procedure is meant to ensure that plastic rotations are restricted to the base region only, which has the advantage of not requiring ductile detailing of the upper part, and of producing a linear drift distribution along the height, thus favouring uniform plastic rotations of the beams in the case of dual systems.

The design values of the shear forces are meant to take into account a possible increase due to the onset of higher vibration modes, after the yielding of the base.

The design shear force is therefore given by the expression

$$V_{Sd} = \epsilon V'_{Sd}$$

where V'_{Sd} is the shear force along the wall, obtained from the analysis, and ϵ is

$$\epsilon = q\left\{\left(\frac{\gamma_{Rd}}{q}\frac{M_{Rd}}{M_{Sd}}\right)^2 + 0{\cdot}1\left[\frac{S_e(T_c)}{S_e(T_1)}\right]^2\right\}^{1/2} \not> q \qquad (2.16)$$

where M_{Rd} is the actual flexural resistance at the base of the wall, as detailed; $S_e(\cdot)$ is the ordinate of the elastic response spectrum for the period T_1 of the fundamental mode of the building and for the period T_c, the corner period of the constant acceleration branch.

For squat walls ($h_w/l_w \leq 2$), the second term in the bracket is omitted.

The capacity design factor γ_{Rd} takes on the values of 1·25 and 1·15 for DCH and DCM, respectively, while for DCL the magnification factor ϵ can be directly assumed to be equal to 1·3.

2.5. Capacity design method in Japan

2.5.1. Introduction

A complete procedure incorporating capacity design concepts and provisions for the design of reinforced concrete frame and wall-frame structures has been developed by the Architectural Institute of Japan (1990). The scope of this procedure is limited to regular building structures with a height not exceeding 45 m. While the objectives of the design are similar to those which are now widely diffused in modern codes, the AIJ procedure for achieving them is, to a great extent, original as will appear from the outline description to follow.

As yet, the AIJ procedure does not have official status in Japan. The seismic action used for the design at the ultimate limit state is taken from the Building Standard Law Enforcement Order (1981), and it is not qualified in terms of average return period and implied amount of structural ductility. The suitability of this action to provide a satisfactory degree of protection is supported by the good overall behaviour exhibited by the building structures in the course of past disastrous events, such as the Kobe earthquake in 1995.

2.5.2. *Outline description of the AIJ procedure*

2.5.2.1. Preferred yield mechanism. For frame structures, the optimum mechanism involves formation of hinges at all beam ends and, at a later stage, at the base sections of the ground-storey columns.

Hinges are allowed in columns at the top sections of the upper storey and in exterior columns when subjected to tension at the ultimate stage. When walls are also present, ductile flexural hinges and uplifting rotation are admitted at their base and yielding in columns is permitted in mid-storeys.

2.5.2.2. Design procedure. The design is carried out in two steps.

- *Step 1.* A static elastic analysis of the structure is performed for the design base shear (typically 25% or 30% of the weight of the building for frame structures and for frames with walls, respectively), distributed along the height proportional to the weight of the storey and to its height from the base.

For buildings more than six storeys high, an additional concentrated force is applied at the top storey, to account for higher modes contribution $F = 0{\cdot}1\ TQ$, where Q is the base shear and T the fundamental period of the building.

The elastic analysis is carried out using realistic secant stiffnesses for the different structural elements, based on their expected deformation level. The bending moments obtained from the analysis at beams ends where hinges are anticipated, can be redistributed to other planned hinges, with a ceiling of 20%. The reinforcement at allowable hinge locations is determined for the redistributed moments.

- *Step 2.* A non-linear push-over analysis is carried out for the same pattern of horizontal loads as in step 1, with the load increasing from zero up to the full formation of the selected mechanism.

During this analysis, only the allowable locations of hinge formation are permitted to yield, while all the other members are modelled as elastic. The resistances of the beams are increased with respect to those previously determined, in order to account for the following sources of overstrength: variability of material properties, contribution of slab reinforcement, possible extra amount of reinforcement actually provided.

At the attainment of the global yield mechanism, the action effects transmitted to the non-yielding members by the yield hinges at their 'overstrength' state are read from the analysis. After further amplification to account for dynamic effects and bi-directional action (bending moments applied to the columns by orthogonal beams simultaneously subjected to seismic effects), the action effects on the columns to be considered for their design are obtained.

2.5.3. *Provisions for ensuring the development of the selected mechanism (capacity design provisions)*

Columns designed according to the elastic analysis in step 1 would not be adequately protected from yielding due to causes of both random (R) and deterministic (D) nature. Causes R can be counteracted by introducing appropriate safety factors, causes D can be removed by correcting the analysis on the basis of the known reality.

(*a*) *Material properties variability (R).* The flexural resistance of beams is essentially governed by steel strength. Based on statistics of observed yield strength (for grade SD 345 steel) a factor of 1·25 is to be applied to the nominal steel strength when the upper bound of flexural resistance is evaluated. Possible overstrength of concrete is disregarded.

(*b*) *Effective slab width (D).* The effective slab width is taken as 0·10 times the clear span of the girder at the design stage. In evaluating the upper bound negative bending moment, the width is doubled, to account for the spreading of yielding to adjacent slab reinforcement.

(*c*) *Amount of longitudinal reinforcement (D).* The push-over analysis is to be performed considering the actual amount of longitudinal reinforcement provided.

(*d*) *Bi-directional earthquake motion (R).* Two-way columns are actually subjected to simultaneous seismic effects in the two orthogonal directions. The worst possible situation for a two-way column would occur in the case of yield mechanisms forming in both directions. This situation is, however, highly improbable, so that AIJ guidelines actually ask for separate and independent analyses for the two directions, but with the bending moments in the columns amplified by a factor of 1·1. Regarding the axial forces, ± 50% of the axial force due to the seismic action acting in the orthogonal direction has to be combined with the maximum/minimum axial force in the main plane of bending.

(*e*) *Dynamic magnification factor (D, R).* Non-linear dynamic analyses conducted on frame and frame–wall structures show that, due to the combined effect of the fundamental mode differing from a straight line, and of higher mode participation, the maximum storey shears are larger than those corresponding to the formation of a beam yield mechanism under a linear distribution of forces.

Denoting the maximum static storey shear by Q_s, and the maximum increment of the storey shear due to the dynamic effects by ΔQ, the amplification factor is defined as

$$\omega = 1 + \frac{\Delta Q}{Q_s}$$

The analyses mentioned above indicate that ΔQ tends to be constant for a given structural type, so that ω becomes smaller for a stronger structure (higher Q_s).

When walls are present in combination with frames, the further problem arises of distributing the increment between the two lateral load resisting systems.

The expressions for ω finally proposed by the AIJ guidelines are

$$\omega_{ci} = 1{\cdot}0 + \frac{\Delta\omega_i}{\phi_o}(\beta_{chi}/\beta_{ci})$$

$$\omega_{wi} = 1{\cdot}0 + \frac{\Delta\omega_i}{\phi_o}(\beta_{whi}/\beta_{wi})$$

where ω_{ci}, ω_{wi} are the amplification factors for columns and walls, respectively, at storey level i; $\Delta\omega_i = 0{\cdot}25$ for $i = 1$; 0·20 for $i \leq n/2$; 0·20 +

$0{\cdot}10\,(i - n/2)$ for $i > n/2$ (n = number of floors); ϕ_o is the ratio between the actual base shear from push-over analysis at the formation of yield mechanism and the design base shear; β_{chi}, β_{whi} are the ratios of the shears carried by column and walls, respectively, to the total shear at storey i, under higher mode oscillations. These ratios may be approximately evaluated by means of the expressions

$$\beta_{chi} = \frac{I_c}{I_c + I_w} \qquad \beta_{whi} = \frac{I_w}{I_c + I_w}$$

where I_c and I_w are the storey stiffness of columns and walls assuming both as fixed at both ends.

2.6. Concluding remarks

The difference between modern seismic design codes regarding the means through which they strive to control inelastic seismic response is one of semantics rather than of essence. First of all, there is a common underlying theme: design is based on the results of a linear elastic analysis for specified lateral forces derived from a design acceleration spectrum, the ordinates of which are obtained by dividing those of the elastic, 5% damped response spectrum by a 'force reduction' or 'behaviour' factor, which is constant for periods in and above the constant spectral acceleration region. Individual members are proportioned for strength on the basis of internal forces, generally computed from this elastic analysis, and are detailed for the peak inelastic deformation demands associated with the design ground motion. The magnitude of inelastic deformation demands on individual members is estimated (and at the same time controlled), if the global displacement ductility demand at the top (which is essentially in one-to-one correspondence with the value of the 'behaviour' of 'force reduction' factor dividing the ordinates of the elastic acceleration spectrum) is distributed as evenly as possible to all members and regions of the structure capable of developing inelastic deformations in a ductile fashion.

Capacity design can be seen as the tool through which the uniform spreading of inelasticity to the ductile regions of the structure is achieved. (Capacity design can also be seen as the means for incorporating within the structure the desired strength hierarchy that eventually dictates the mode of inelastic action and failure.) The prescriptive detailing rules through which the achievement of the desired member and region ductility supply is pursued, refer to the quantity, shape and spacing of transverse reinforcement, to the arrangement and the limiting values (minimum, and for the tensile reinforcement also maximum) of the longitudinal steel and to the extent of the member volume over which these rules apply. These prescriptive rules differ from code to code, mainly due to differences in the value of the global displacement ductility factor implicit in the code.

The value of the desired ductility supply also affects the extent and the way in which different codes apply capacity design to control the distribution of inelastic deformation demands: typically codes provide for more than one strength–ductility combination, allowing ductility to be traded against strength, depending on the seismicity of the region and on the particular features of the structure. Eurocode 8 does this through the three ductility classes (DC), low (L), medium (M) and high (H). In the USA the UBC and the NEHRP recommendations adopt the ACI-318 categorization of RC structures into 'ordinary moment frames', 'intermediate moment frames' and 'special moment frames', and specify the application of each type depending on the seismicity of the region (the UBC) or on the desired 'seismic performance' (the NEHRP recommendations). The New Zealand code allows the designer to select a global displacement ductility factor μ_δ

less than the value of 6 which corresponds to fully ductile frames. Design lateral forces are then increased, as the design spectrum is a function of μ_δ. In all these cases, capacity design rules are relaxed, as the desired ductility level decreases.

Capacity design rules for columns in bending and for beams and columns in shear aim at avoiding overstrength in the ductile modes of behaviour and (possible) failure, such as those (mainly of beams) in flexure, relative to the more brittle ones, such as those of all elements in shear and of columns in flexure. Such overstrengths may occur if the resistance of the more ductile modes is controlled by gravity loads or minimum reinforcement, while that of more brittle modes is controlled by the design seismic action.

In lower ductility structures design seismic internal forces are in the order of 50% or more of those resulting from purely elastic response to the design ground motion. For such high design seismic forces, it is expected that the seismic action will control proportioning of every member against all failure modes and that undesirable overstrengths will not take place. Moreover, the demand values of the member ductility factors associated with the low global displacement ductility factors of low ductility structures, are relatively low, even if inelastic deformation demands are not uniformly distributed within the structure. These low local ductility demands can be easily accommodated through detailing of beams and columns which is only slightly more demanding than that required in non-earthquake resistant structures. Accordingly, capacity design requirements can be waived, especially as their application greatly complicates the entire phase of proportioning/detailed design of members. For example, Eurocode 8 and the USA standards (essentially the ACI–318 code) impose no capacity design requirement whatsoever on structures belonging to the class of DCL structures or of 'ordinary moment frames', respectively. USA standards waive the columns of 'intermediate moment frames' from capacity design in bending. For 'special moment frames' this latter aspect of capacity design is applied, not through a multiplication of column design moments M_{Sc} from the analysis by the capacity design magnification factor $\gamma_{\mathrm{Rd}}/\delta$ or $1 + \gamma_{\mathrm{Rd}} - \delta$ of Eurocode 8 or $\varphi_{\mathrm{oj}} = \Sigma M_{\mathrm{bo}}/\Sigma M_{\mathrm{bE}}$ of the New Zealand code, but as an *a posteriori* check that the sum of column uniaxial moment capacities ΣM_{Rc} exceeds that of the beams ΣM_{Rb} with an overstrength margin of 20%. If this check is not satisfied in one of the two horizontal directions at a single level of a column of such a frame, the contribution of that column to the lateral strength and stiffness of the frame must be disregarded, and the column has to be proportioned for gravity loads alone, while still respecting all the requirements for minimum longitudinal and transverse reinforcement of special moment frames, so that it can sustain the ductility demands imposed by the rest of the lateral force resisting system, the displacements of which it shares. (Obviously, considering the strength and stiffness of a column for one set of actions — the gravity actions — and disregarding it for another, is very inconvenient from the modelling and analysis point of view.)

The Eurocode 8 provisions for the capacity design of DCM and DCH columns in bending differ from those of the other codes in two respects. First, in that the magnification of column moments is applied simultaneously in both principal directions of column bending with the appropriate magnification factors, leading to proportioning of column sections for biaxial bending. Secondly, in the relaxation allowed when the beam moment capacity ΣM_{Rb} is not exhausted by the corresponding seismic action effects according to the analysis ΣM_{Sb}. In this respect, the New Zealand approach of full capacity design with so many different special cases and values (dual systems versus pure frames, one-way versus two-way columns, columns dominated by frame versus cantilever action, ductile frames versus frames of limited ductility with their many special cases, etc.), is computationally inconvenient. (Indeed, the New Zealand approach is intentionally more hand-calculation oriented than any other modern code approach.) On the contrary, the Eurocode 8 approach of relaxed capacity design applies uniformly to all types of structural systems and columns (gravity or seismic action dominated ones, dual systems or frames, frame or cantilever action dominated columns, etc.), automatically taking into account the special cases and the effect of the various parameters handled by other standards through special rules. Moreover, it applies simultaneously to both horizontal directions of bending, giving the appropriate weight to the principal bending direction. Therefore, this general capacity design approach is computationally very convenient.

The New Zealand capacity design approach for columns in bending shares some common features with that of the 1990 AIJ guidelines. In addition to including a multiplicative factor of 1·1 to 1·2 on column uniaxial moments to account in approximation for column biaxial bending, these standards provide for an additional multiplicative factor ω on column moments, to account for higher mode dynamic effects that might cause the ratio of column moments above and below a joint (or at the top and bottom of a storey) to deviate significantly from that obtained from a first-mode dominated analysis. Therefore, these two standards, which do not allow any capacity design relaxation for non-seismic dominated beams, provide almost full protection of columns from plastic hinging.

Code capacity design rules for shear are also a function of the required ductility level. For beams, Eurocode 8 requires capacity design in shear only for DCH structures. The USA standards do so for the beams and columns of 'special moment frames', while for those of 'intermediate moment frames' they allow instead the use of the seismic action effects from the analysis multiplied by 2·0. Recognizing that, due to the relaxation of column capacity design, there is a possibility of plastic hinge development in the columns, Eurocode 8 imposes capacity design of DCH and DCM columns in shear assuming column yielding (and at overstrength) at top and bottom, regardless of what the beams are doing there. On the contrary, the New Zealand code, the Japanese Law Enforcement Order and the USA standards (whenever they require column capacity design in shear) allow the column design shear force to be based on the beam flexural capacities (at overstrength) instead of those of the column itself, if it is the beams rather than the column that yield first around a joint.

When it allows the designer to select a global displacement ductility factor (less than the value of 6 corresponding to fully ductile frames), the New Zealand standard also allows some (or even all) of the columns (especially internal ones) to develop plastic hinges. The ductility factor to be used for the design (typically between 1·25 and 3 in these 'limited ductility' frames) is a function of the fraction of storey shear carried by those columns which are protected from plastic hinge development through the application of the associated capacity design rule. This rule, however, is applied with lower dynamic magnification factors than in the case of fully ductile frames, namely with $\omega = 1{\cdot}1$ or $1{\cdot}3$ for 'one-way' and 'two-way' columns respectively, regardless of the value of the fundamental period. Capacity design of beams and columns in shear is essentially, as in fully ductile frames, dependent on whether the member is expected to develop plastic hinges or not. In addition to the relaxation of capacity design in flexure for the columns, column detailing requirements are also relaxed, in return for the increase in design forces: in columns expected to avoid plastic hinge development and remain elastic, stirrup spacing for the prevention of bar buckling is as in detailing for normal (i.e. non-seismic) actions, while if the design global ductility factor μ_δ is less than 3, the confining steel in columns allowed to develop plastic hinges is computed on the basis of a curvature ductility factor μ_φ of 10, instead of 20 in plastic hinges of columns of fully ductile frames.

Clearly, the New Zealand design provisions for alternative strength–ductility combinations give considerable flexibility and freedom to the designer, without unduly complicating the design process. In comparison, the USA and the European standards, which provide for a few 'discrete' strength–ductility combinations, each with its own well defined rules for member proportioning and detailing, are most convenient for computational implementation and routine application, but significantly limit the choices available to the designer.

Regarding capacity design of walls in bending and in shear, both the New Zealand and the European standards try to limit inelastic action to the base region of the wall by proportioning all sections above the base for a linear envelope of moments, defined at the base by the design moment there from the analysis M_{Sw} according to Eurocode 8, or by the flexural capacity of the base section as designed and detailed and at overstrength, $\gamma_{Rd}M_{Rw}$, according to the New Zealand code. Both these codes magnify design shears depending on flexural overstrength ratio at the as-detailed wall base, $\gamma_{Rd}M_{Rw}/M_{Sw}$, and accounting semi-empirically for higher mode inelastic effects on wall shear forces (but in a slightly different way).

USA standards do not include specific provisions for capacity design of walls in flexure or in shear, relying on their high stiffness and on the capacity design of columns for avoidance of soft-storey formation and for the control of inelastic action.

3. Reliability-based system analysis

3.1. Introduction

The previous two chapters have already made it clear that in order to design a structure which is capable of providing specified performances when acted upon by a seismic type of excitation, it is essential to deal with the behaviour of the structural system in its entirety. Building structures, in particular, are characterized by a pronounced degree of seriality in the way in which their sub-parts are arranged (e.g. consecutive storeys, beam–column–joint subassemblages) and in the way in which different resisting mechanisms coexist within the elements (e.g. flexure and shear), so that it is immediately obvious that malfunctioning of even a few of the sub-systems/mechanisms directly affects the response of the whole system. The consequence can either be outright failure (as, for example, when a storey sway mechanism is formed), or a decreased efficiency of the system in dissipating part of the energy imparted by the soil, with possible accelerated degradation of the mechanical properties (e.g. shear and bond failures, severe cracking of joints, etc.).

Within the scope of this Design Guide, attention is restricted to the provisions which are needed to ensure that a (bare) RC framed structure will respond to the design seismic action with inelastic flexural behaviour concentrated essentially in the beams, all other forms of inelastic behaviour being excluded. The problem under consideration is therefore only a part of the whole: it is, however, the most difficult one since it implies control of the dynamic response of the entire structure, while all other problems are solvable locally as a matter of relative strength between desired and undesired mechanisms.

The solution given to this problem by some of the major international codes in their latest versions has been presented in Chapter 2. Over the years, starting from the early 1970s, the so-called capacity design (CD) procedures have evolved from small sets of rules to rather sophisticated processes, as is particularly the case for New Zealand. Although the single steps in these procedures are logical and reflect good engineering sense, unsatisfactory aspects remain. Firstly, the serious lack of quantitative assessments of the proposed CD procedures may leave one with the impression that, at least for some of them, the level of protection they actually provide is unknown and uncontrolled, possibly on average higher than in the intentions of the code-makers themselves. From a more conceptual point of view, one is left wondering if, in place of the various reasonable but entirely subjective assumptions adopted (for example in the establishment of the column axial forces based on the beams of the upper floors developing their overstrength, etc.), recourse to a more objective formal approach would be feasible, if for no other reason than that of validating the above rules, but also to extend the scope of the present procedures to cover both different types of structures (CD procedures are inherently typology dependent) and, perhaps more interestingly, performance requirements which differ from the usual ones.

This possibility clearly exists, in principle, and is given by the methods of reliability analysis of structural systems. It must be admitted, however, that these methods find here their most challenging and not entirely developed field of application: because system response is dynamic and highly non-linear, and because both the input and the system itself need to be realistically treated as random quantities. Since the reach of these methods barely extends to coverage of the complexity of the problem at hand, no miraculous, all-encompassing solutions can be expected, but rather indications of the relevance of the different sources of variability and uncertainty, and of the interplay among them. Deterministic calibrations using refined mechanical models to investigate specific problems will still be necessary.

It is important to note that capacity design is actually a package of measures, some of which are intended to counter real physical variabilities and subjective uncertainties, while some others are corrective measures meant to redress 'inaccuracies' arising from the design procedures. This is particularly the case for the ω-factors in the New Zealand code, and for the different ways in which slab reinforcement is counted (with good reason) when dimensioning for the negative moment resistance of a beam and when assessing its potential overstrength in the presence of large inelastic rotations.

Finally, one should take into account the fact that CD procedures combine with other fixed provisions meant to ensure ductile behaviour, like minimum reinforcement for positive moment in beams, minimum percentage of reinforcement of columns, etc., and that the relative importance of these latter factors is very variable between structures and within the same structure at different places. Although the beneficial influence (and hence the necessity) of the minimum requirements is beyond question, it is a fact that they make the assessment of the effectiveness of the CD procedures *per se* more complex and more case dependent.

A complete probabilistic approach to the calibration of CD procedures should be able to deal with the following aspects

(*a*) the randomness of the seismic motion
(*b*) the variability of the mechanical properties of the material
(*c*) the variability in the response due to the models used for the analysis
(*d*) the uncertainties with regard to the resistance of the elements/mechanisms.

Only aspects (*a*) and (*b*) are considered in the application presented in this chapter. Model uncertainty (aspect (*c*)) could easily be included in the procedure, if data were available on the scatter produced by the simplified (but assumed to be 'true' in the mean) method of structural analysis that has been adopted. Lack of explicit recognition of this source of scatter, however, may be thought to be covered implicitly by a somewhat larger variability attributed to sources (*a*) and (*b*).

Finally, the analysis presented in the following sections terminates with the determination of the second moment properties of the response variables of interest, like displacements, drifts, local ductilities, etc. A full risk analysis of the structures (i.e. determination of the probability of failure, expressed for example as the probability that the point representative of the random vector response process outcrosses, in a given time interval, a random safe domain defined in the same space) would be feasible, under suitable simplifying assumptions, although at considerable computational expense. It has not been pursued, however, believing it to be beyond the scope of the investigation. As it stands, the results can be presented in the same format as in all other calibration studies: the envelope of the maximum (fractile) values of the responses, against the corresponding envelope of the (fractile) values of the capacities.

The technique used for the probabilistic analysis of the response is the equivalent stochastic linearization (ESL), well known for many years, and extended here to treat the specifics of the problem examined. The technique and its extension are briefly described in the following section.

3.2. Objective and scope

The suitability of ESL for the task discussed in Section 3.1 has been examined in a number of papers: Colangelo *et al.*, 1994, Pinto *et al.*, 1995, Colangelo *et al.*, 1995. The method itself has undergone significant developments over the years. Starting from the initial application (Atalik and Uktu, 1976; Wen, 1976), improvements have regarded the introduction of the degradation of the mechanical properties, stiffness and strength, as a function of the dissipated energy or of the displacements (Baber and Wen, 1981), the extension to biaxial bending (Park *et al.*, 1986), and the analysis of the sensitivity of the response statistics to the values of the structural parameters (Sues and Wen, 1985), a result which opens the way to a full stochastic analysis of the structure.

Two main limitations remain: the approximate character, which is intrinsic to the nature of the method and is therefore insurmountable, and the modelling capabilities, which are practically restricted to a class of differential constitutive models and cannot in all cases accurately reproduce the actual behaviour. It can be validly argued, however, that neither limitation interferes with the specific use of the method; what is required for calibration is not a high degree of precision in absolute terms, but an acceptable accuracy in quantifying the differences of the results due to variations in the parameters.

A first application of ESL to check its suitability for the calibration of CD procedures has been presented in Colangelo *et al.*, 1994, where a small sample of frame structures, designed for different combinations of q and CD factors, was considered. In this first application, the seismic motion was the only source of variability, while the structure was taken as deterministic. Furthermore, the frames were designed and checked for seismic action only, assuming its effect to be predominant over that due to vertical loads. The effect of stiffness and strength degradation on the required ductility, and hence on the values of the CD factors to keep the demand constant, have been examined in Pinto *et al.*, 1995, using the same population of structures considered in the previous study and the same assumptions of deterministic structure and predominant seismic action. Randomness of the mechanical properties, and in particular of the flexural strengths, has been introduced in Colangelo *et al.*, 1995, again for the same population and in the absence of vertical loads: the results indicated that structural variability plays a considerable role in the quantification of the CD factors.

The present study extends the previous ones, essentially along two lines. First, the effect of vertical permanent loads is introduced. This has required a modification of the method to include unsymmetrical behavioural laws. The loss of symmetry involves a penalty in terms of computational efficiency, but is a necessary feature if one wants to include large permanent loads in the calibration cases. Secondly, the structural typologies have been extended to include frames showing irregular patterns along the height, as well as frames of industrial buildings, designed to carry large service loads. Since Eurocode 8 contains special provisions of q and CD factor values for the situations indicated above, the study offers a (limited) contribution to the validation of these provisions.

With the present study, the potential of ESL for the aim under consideration can be said to have been sufficiently explored to confirm its usability as a theoretically appropriate tool for a systematic support of code provisions.

3.3. Stochastic linearization

A structure subjected to a zero mean process, as the seismic input is usually modelled, may have a non-zero mean response as a consequence of two factors

(*a*) the presence of vertical loads
(*b*) the unsymmetric force–deformation relationship of the structure,

which, in turn, is due to the fact that the design is usually made to satisfy a combination of vertical and seismic forces which do not yield a symmetrically behaving structure.

To consider the response as a zero mean process, by disregarding the presence of the vertical loads and assuming symmetrical behavioural laws, may still lead to meaningful results in some cases, but cannot serve the specific purposes of the present work because

(*a*) it cannot distinguish between cases of varying relative importance of vertical and seismic forces
(*b*) it cannot account for the unsymmetrical flexural strength of the beams.

The simplification of a zero mean response process has therefore to be abandoned. Solutions for the case of hysteretic structures subjected to non-zero mean input processes are already available (Baber, 1984, 1986) for the case of symmetric structural behaviour: in the following sections the generalization is given to cover the unsymmetric cases as well.

3.3.1. Frames with unsymmetrical behaviour

The system under consideration is made up of linear elastic elements with inelastic hinges at their extremities. Denoting by $\boldsymbol{u}$ the vector of nodal displacements and by $\boldsymbol{h}$ the vector of bending moments in the hinges, the equation of dynamic equilibrium of the system can be expressed as

$$\mathbf{M}\ddot{\boldsymbol{u}} + \mathbf{C}\dot{\boldsymbol{u}} + \mathbf{R}\boldsymbol{h} = \boldsymbol{f}(t) \qquad (3.1)$$

where $\mathbf{M}$ and $\mathbf{C}$ are the mass and damping matrices, respectively, $\boldsymbol{f}(t)$ is the vector of structural actions and $\mathbf{R}$ is an appropriate location matrix.

3.3.1.1. Moment–rotation law. The moment–rotation law of the hysteretic hinge is idealized by means of the smooth differential model proposed by Bouc and later generalized by Wen (1976). As is well known, the model is quite versatile, being capable of reproducing both hysteretic and conservative laws, as well as softening and hardening behaviours, through an appropriate definition of its parameters. In addition, closed-form solutions are available for the stochastic linearization of softening hysteretic systems. In the present chapter an extended Bouc-Wen equation is proposed to model different positive and negative yielding moments of member sections. Following Wen, 1976, each moment h_i is written as the sum of a linear-elastic contribution and a separate, purely hysteretic component

$$h_i = \alpha_i k_{\theta i}\theta_i + (1 - \alpha_i)k_{zi}z_i \qquad (3.2)$$

where α_i measures the relative importance between the elastic term and the hysteretic one. Herein θ_i denotes the plastic hinge rotation and z_i is a history-dependent auxiliary variable which results from integrating the following differential equation

$$\dot{z}_i = A_i\dot{\theta}_i - \gamma_i\dot{\theta}_i|z_i|^{n_i} - \beta_i|\dot{\theta}_i||z_i|^{n_i-1}z_i - \delta_i\dot{\theta}_i|z_i|^{n_i-1}z_i \qquad (3.3)$$

A_i, γ_i, β_i and n_i are parameters governing the hysteresis shape, as shown in detail, for example in Wen, 1976, while the additional parameter δ_i determines asymmetric yielding levels for hinge i. In fact, once equation (3.3) is written as

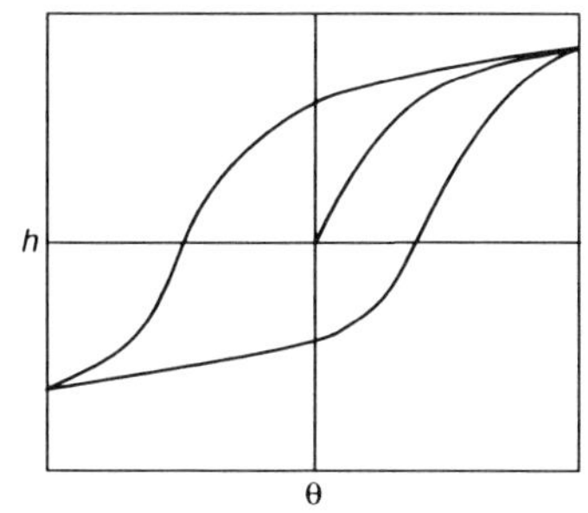

Fig. 3.1. Sample of an asymmetric cycle of the modified Bouc–Wen law

$$\frac{dz_i}{d\theta_i} = A_i - |z_i|^{n_i}[\gamma_i + \beta_i \text{sign}(z_i d\theta_i) + \delta_i \text{sign}(z_i)]$$

it appears clearly that $|z_i|$ tends towards the extreme values

$$|z_i| \rightarrow \left(\frac{A_i}{\gamma_i + \beta_i \pm \delta_i}\right)^{1/n_i}$$

provided that $A_i > 0$ and $\gamma_i + \beta_i \pm \delta_i > 0$. The value of δ_i regulates the difference between the positive yielding moment and the negative one; its sign determines which resistance is stronger. As an example, an asymmetric cycle is shown in Fig. 3.1.

3.3.1.2. First moments equations. The moments in the springs must be in equilibrium with those at the end of the elastic elements; hence they have to satisfy the equations

$$\boldsymbol{h} = \mathbf{K}_{hu}\boldsymbol{u} - \mathbf{K}_{h\theta}\boldsymbol{\theta} \tag{3.4}$$

where $\mathbf{K}_{hu}$ and $\mathbf{K}_{h\theta}$ are appropriate matrices containing stiffness terms of the elastic beam elements and $\boldsymbol{\theta}$ is the vector of plastic rotations in the hinges. In matrix notation the equations (3.2) and (3.3) can be expressed as

$$\boldsymbol{h} = \mathbf{A}\boldsymbol{\theta} + \mathbf{B}\boldsymbol{z}$$

$$\dot{\boldsymbol{z}} = \boldsymbol{g}(\boldsymbol{z}, \dot{\boldsymbol{\theta}})$$

and, by making use of the former in equations (3.1) and (3.4) the following system is obtained

$$\begin{aligned} \mathbf{M}\ddot{\boldsymbol{u}} + \mathbf{C}\dot{\boldsymbol{u}} + \mathbf{R}\mathbf{K}_{hu}\boldsymbol{u} - \mathbf{R}\mathbf{K}_{h\theta}\boldsymbol{\theta} &= \boldsymbol{f}(t) \\ \mathbf{K}_{hu}\boldsymbol{u} - (\mathbf{K}_{h\theta} + \mathbf{A})\boldsymbol{\theta} &= \mathbf{B}\boldsymbol{z} \\ \dot{\boldsymbol{z}} &= \boldsymbol{g}(\boldsymbol{z}, \dot{\boldsymbol{\theta}}) \end{aligned} \tag{3.5}$$

Taking now the expectation of equation (3.5), and noting that in the stationary case considered here the averages of the process do not vary with time, one gets

$$\begin{aligned} \mathbf{R}\mathbf{K}_{hu}\boldsymbol{\mu}_u - \mathbf{R}\mathbf{K}_{h\theta}\boldsymbol{\mu}_\theta &= \boldsymbol{\mu}_f \\ \mathbf{K}_{hu}\boldsymbol{\mu}_u - (\mathbf{K}_{h\theta} + \mathbf{A})\boldsymbol{\mu}_\theta &= \mathbf{B}\boldsymbol{\mu}_z \\ \mathrm{E}\{\boldsymbol{g}(\boldsymbol{z}, \dot{\boldsymbol{\theta}})\} &= 0 \end{aligned} \tag{3.6}$$

where $\boldsymbol{\mu}_x$ denotes the expected value of the random process $\boldsymbol{x}$. The expected value of $\boldsymbol{g}(\cdot)$ is to be evaluated with reference to equation (3.3), and the results can be expressed as

$$E\{g_i(z_i, \dot{\theta}_i)\} = -\gamma_i c_{01i} - \beta_i c_{02i} - \delta_i c_{03i} \tag{3.7}$$

with the coefficients c_{0ki} ($k = 1, 2, 3$) being functions of the mean values and of the covariance matrix of z_i, $\dot{\theta}_i$. The explicit expressions of c_{01} and c_{02} are given in Baber, 1984, for the non-stationary case; the corresponding expressions for the stationary case are obtained by putting to zero the time derivatives of the mean values. The expression for c_{03} is given in Colangelo *et al.*, 1996.

3.3.1.3. Second moments equations. If the two first equations of the set (3.6) are subtracted from the corresponding ones in the set (3.5), and $\boldsymbol{u}_0$, $\boldsymbol{\theta}_0$. $\boldsymbol{f}_0$, z_0 are used to indicate the difference between each quantity and its corresponding mean value, so that all of them represent zero–mean processes, the following set is obtained

$$\begin{aligned} \mathbf{M}\ddot{\boldsymbol{u}}_0 + \mathbf{C}\dot{\boldsymbol{u}}_0 + \mathbf{RK}_{hu}\boldsymbol{u}_0 - \mathbf{RK}_{h\theta}\boldsymbol{\theta}_0 &= \boldsymbol{f}_0(t) \\ \mathbf{K}_{hu}\boldsymbol{u}_0 - (\mathbf{K}_{h\theta} + \mathbf{A})\boldsymbol{\theta}_0 &= \mathbf{B}z_0 \\ \dot{z}_0 &= \boldsymbol{g}(z_0 + \dot{\boldsymbol{\mu}}_z, \dot{\boldsymbol{\theta}}_0) \end{aligned} \tag{3.8}$$

The system (3.8) is formally the same as for the case of zero–mean processes; the last equation is the only non-linear one in the set; the linearized form can be expressed as

$$\dot{z}_0 = \mathbf{H}_z z_0 + \mathbf{H}_\theta \dot{\boldsymbol{\theta}}_0 \tag{3.9}$$

where $\mathbf{H}_z$ and $\mathbf{H}_\theta$ are diagonal matrices made up of the expected values of the derivatives of the function $\boldsymbol{g}(\cdot)$

$$\begin{aligned} \mathbf{H}'_z &= E\left\{\frac{\partial \boldsymbol{g}}{\partial z_0}\right\} = \mathrm{diag}(-\gamma_i c_{11i} - \beta_i c_{12i} - \delta_i c_{13i}) \\ \mathbf{H}_\theta &= E\left\{\frac{\partial \boldsymbol{g}}{\partial \boldsymbol{\theta}_0}\right\} = \mathrm{diag}(A_i - \gamma_i c_{21i} - \beta_i c_{22i} - \delta_i c_{23i}) \end{aligned} \tag{3.10}$$

A common assumption in earthquake engineering is to model the seismic motion as the output of a linear filter subjected to a white noise process. If $\mathbf{T}$ denotes the matrix of the rigid base translation and $\mathbf{F}$ the matrix of the filter coefficients, one may write the zero–mean part of the vector of the nodal forces as

$$\boldsymbol{f}_0(t) = \mathbf{MT}\boldsymbol{y}(t) \qquad \dot{\boldsymbol{y}} = \mathbf{F}\boldsymbol{y} + \boldsymbol{w} \tag{3.11}$$

Upon substitution of equation (3.9) into the last of equation (3.8), writing the first of equation (3.8) in the form of a first order equation, and introducing the vector $\boldsymbol{x}$ of the unknowns $\boldsymbol{x} = \{\boldsymbol{u}'_0\ \dot{\boldsymbol{u}}'_0\ \boldsymbol{\theta}'_0\ \boldsymbol{y}'\}'$, the linearized set becomes

$$\dot{\boldsymbol{x}} = \mathbf{G}\boldsymbol{x} + \boldsymbol{c} \tag{3.12}$$

where matrix $\mathbf{G}$ depends on the characteristics of the structure, on those of the hysteretic springs and on the filter parameters

$$\mathbf{G} = \begin{bmatrix} 0 & \mathbf{I} & 0 & 0 \\ -\mathbf{M}^{-1}\mathbf{RK}_{hu} & -\mathbf{M}^{-1}\mathbf{C} & \mathbf{M}^{-1}\mathbf{RK}_{h\theta} & \mathbf{T} \\ -\mathbf{EH}_z\mathbf{B}^{-1}\mathbf{K}_{hu} & \mathbf{EB}^{-1}\mathbf{K}_{hu} & \mathbf{EH}_z\mathbf{B}^{-1}(\mathbf{K}_{h\theta} + \mathbf{A}) & 0 \\ 0 & 0 & 0 & \mathbf{F} \end{bmatrix}$$

with $\mathbf{E} = [\mathbf{H}_\theta + \mathbf{B}^{-1}(\mathbf{K}_{h\theta} + \mathbf{A})]^{-1}$. The right hand vector $\boldsymbol{c}$ is a function of the input process $\boldsymbol{w}$: $\boldsymbol{c} = \{0' \; 0' \; 0' \; \boldsymbol{w}'\}'$. Equation (3.12) provides the basis for the second moment equation

$$\dot{\mathbf{V}} = \mathbf{GV} + \mathbf{VG}' + \mathbf{D} = 0 \tag{3.13}$$

in which $\mathbf{V}$ is the covariance matrix of $\boldsymbol{x}$ and $\mathbf{D}\delta(t) = E[\boldsymbol{cc}']$.

The solution of equation (3.13) gives the variance of the response, while the mean value is given by the solution of the first two equations (3.6), together with equation (3.7). Since some of the coefficients in equation (3.13) and (3.7) depend on the means and variances of the response, i.e. on the solution, an iterative approach is required.

3.3.2. *Variability of mechanical parameters of the structure*

The square root of the diagonal terms of matrix $\mathbf{V}$ gives the standard deviation of the response components. The peak values of the process (for a given probability of exceedance) can be put in the usual form

$$x_k = \mu_x \pm p_k \sigma_x \tag{3.14}$$

where p_k = peak factor dependent on the probability value, on the power density function of the process and on its duration, and the + or − sign depends on the sign of the mean μ_x. As a rough approximation, p_k can be assumed as a constant for all the components of the process.

The standard deviations of the response are clearly functions of the mechanical parameters of the structure. If $\boldsymbol{q} = \{q_i\}$ $(i = 1, 2, \ldots, n_q)$ is the vector containing these parameters, the fractiles of the response can be expanded as function of $\boldsymbol{q}$ in the form

$$x_k = x_{k0} + \sum_i \frac{\partial x_k}{\partial q_i}(q_i - q_{0i}) + \frac{1}{2}\sum_{i,j}\frac{\partial^2 x_k}{\partial q_i \partial q_j}(q_i - q_{0i})(q_j - q_{0j}) + \ldots \tag{3.15}$$

where $x_{k0} = x_k(\boldsymbol{q}_0)$ indicates the value of x_k corresponding to the reference $\boldsymbol{q}_0$.

If now the parameters $\boldsymbol{q}$ are considered to be random, the fractile values x_k also become random quantities, related to $\boldsymbol{q}$ by equation (3.15). Assuming $\boldsymbol{q}_0$ to represent the mean value of the random Gaussian vector $\boldsymbol{q}$, the mean value and the variance of x_k are easily derived from equation (3.15) knowing the covariance matrix of $\boldsymbol{q}$, $\sigma_{ij} = E\{(q_i - q_{0i})(q_j - q_{0j})\}$

$$E\{x_k\} = x_{k0} + \frac{1}{2}\sum_{i,j}\frac{\partial^2 x_k}{\partial q_i \partial q_j}\sigma_{ij}. \tag{3.16}$$

$$\mathrm{Var}\{x_k\} = \sum_{i,j}\frac{\partial x_k}{\partial q_i}\frac{\partial x_k}{\partial q_j}\sigma_{ij} + \frac{1}{4}\sum_{i,j,h,l}\frac{\partial^2 x_k}{\partial q_i \partial q_j}\frac{\partial^2 x_k}{\partial q_h \partial q_l}(\sigma_{ih}\sigma_{jl} + \sigma_{il}\sigma_{jh}) \tag{3.17}$$

From the mean value and the variance of x_k a convenient fractile of x_k: x_{kk}, can be formed, for example, by taking the mean plus one standard deviation

$$x_{kk} = |E\{x_k\}| + \surd(\mathrm{Var}\{x_k\}) \tag{3.18}$$

In the applications to follow, equation (3.18) has been assumed as measure of the maximum of the x component of the response process, also taking account of the randomness of the mechanical parameters. This simplified formulation, based on first and second moments only, has been preferred over possible alternatives, whose greater accuracy is considered to be purely mathematical, since the actual distribution of the process is not known owing to the linearization introduced.

Member cross-sections

Storey level	Outer column	Inner column	Beam	
1, 2, 3, 4	0·45×0·45	0·50×0·50	0·30×0·50	Four-storey frames
7, 8	0·45×0·45	0·55×0·55	0·35×0·50	Eight-storey frames
5, 6	0·50×0·50	0·60×0·60	0·35×0·60	
3, 4	0·55×0·55	0·65×0·65	0·35×0·70	
1, 2	0·60×0·60	0·70×0·70	0·35×0·70	
3	0·40×0·55	0·40×0·55	0·40×1·10	Three-storey frames
2	0·50×0·65	0·50×0·65	0·50×1·20	
1	0·50×0·75	0·50×0·75	0·50×1·20	

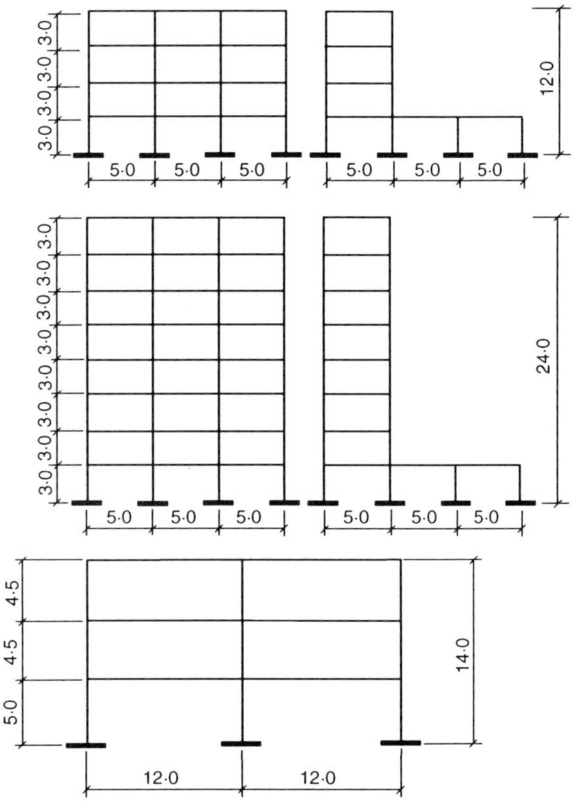

Fig. 3.2. Geometry of frames analysed

Equation (3.17) contains the first and second derivatives of x_k: the number of these latter is $n_q(n_q+1)/2$, i.e. grows with the square of n_q, and can hence lead to considerable computational demands. In the applications presented herein, the mixed derivatives have been found to contribute insignificantly in a number of test cases, and have therefore partially been omitted in the actual examples.

The theory exposed in this section has been implemented in a computer program for the stochastic analysis of non-linear framed structures (Colangelo, 1994).

3.4. Applications

3.4.1. Structures considered

The structures considered consist of planar, reinforced concrete frames, 3, 4 and 8 storeys high, whose geometry is shown in Fig. 3.2. The 4- and 8-storey frames have also been analysed for an irregular configuration, in which only one bay continues for the full height, while the others remain one or two storeys high. An important difference between the 3-storey frame and the others is that the former is meant to carry relatively large vertical loads, while in the others the effects of vertical load are of secondary importance with respect to the seismic action.

3.4.2. Design criteria

All frames have been designed according to the provisions of Eurocode 8, using multi-modal analysis and the design response spectrum valid for intermediate soil types (B). The peak ground acceleration has been assumed to be equal to $a_g = 0{\cdot}35\,g$; the designs have been carried out for the mean ductility class (DCM), for which the behaviour factor is $q = 3{\cdot}75$ in the case of regular, and $q = 0{\cdot}8 \times 3{\cdot}75 = 3{\cdot}0$ for structures which are irregular along their height.

The seismic load combination given in EC8, which includes, besides the seismic action, all permanent and a fraction of the variable loads, has been

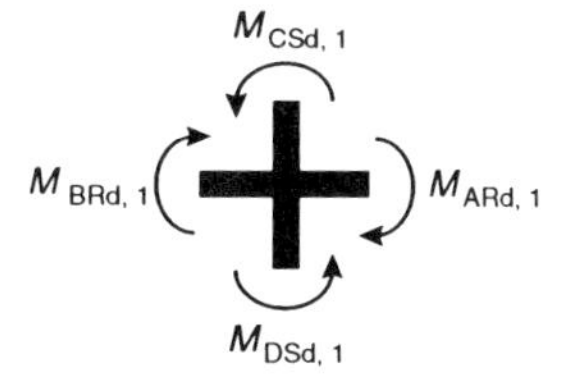

Fig. 3.3 Signs of seismic moments in a generic node

adopted for dimensioning the elements. In parallel, for the purpose of assessing the effect of the vertical loads in the provisions of capacity design, the same frames have also been designed and analysed considering the seismic action only.

The flexural strength of the beams has been assumed as coincident with the requirements of the analysis; further, in any section the minimum resisting moment is never less than half of the maximum with the opposite sign.

The flexural strength of the columns has been obtained by amplifying the values from the analysis by the CD factors α_{CD}

$$\alpha_{CD,i} = \gamma_{Rd}\frac{M_{ARd,i} + M_{BRd,i}}{M_{CSd,i} + M_{DSd,i}}$$

with γ_{Rd} = 1·20 for ductility class M. M_{ARd} and M_{BRd} are the resisting moments of the beams, M_{CSd}, M_{DSd} the values given by the analysis for the columns. The index i identifies the two cases of opposite sign of the seismic action. The signs of the moments for the condition $i = 1$ is shown in Fig. 3.3.

In EC8 the relative importance of gravitational loads with respect to seismic loads is measured through the so-called 'moment reversal factor' δ_i

$$\delta_i = \frac{M_{ASd,i} + M_{BSd,i}}{M_{ARd,i} + M_{BRd,i}}$$

This factor is lower the larger are the overstrengths of the beams with respect to the requirements of the analysis (according to the design procedure followed, δ_i is lower the larger is the ratio between $M_{BRd,1}$ and $M_{BSd,1}$). The design bending moment in the columns $M_{Sd,i,CD}$, to be calculated separately for the two signs of the seismic action, is finally given by

$$M_{Sd,iCD} = |1 + (\alpha_{CD,i} - 1)\delta_i|M_{Sd,i} \leq qM_{Sd,i}$$

Symmetric reinforcement has been adopted for the columns, which implies designing for the larger of the two moments $M_{SD,i,CD}$. At the base section of the ground floor the amplifying factor evaluated for the top section is applied.

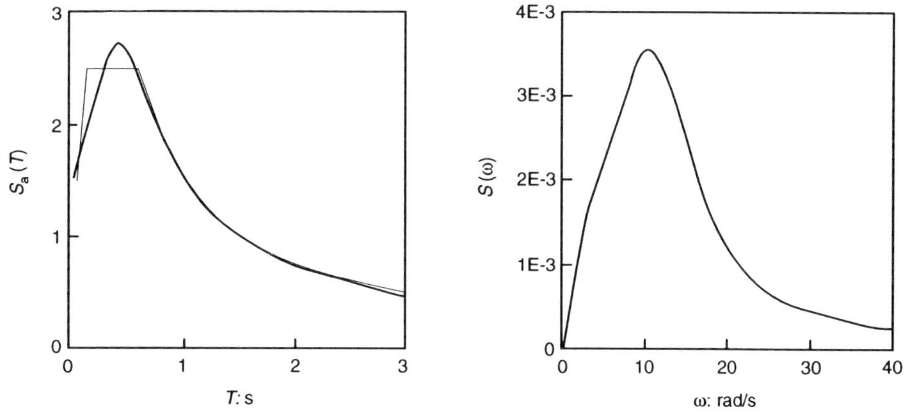

Fig. 3.4. Response $S_a(T)$ and power $S(\omega)$ spectra of filtered white noise compared with EC8 soil M response spectrum

3.4.3. *Reliability analysis*

The ESL procedure has been applied to the probabilistic evaluation of the response of the frames designed as described above.

As regards the input, the filter is composed of two serial linear filters of second order whose parameters have been adjusted so as to match as closely as possible the power spectral density underlying the response spectrum adopted for the design, with the ordinates interpreted as 50% fractile (Fig. 3.4).

The randomness in the structural parameters has been limited to the flexural strength of the plastic hinges. The spatial modelling of randomness is as follows: the strengths of all hinges belonging to the beams of a floor, as well as those of the column hinges located immediately above and below the same floor are assumed each to depend on a single random variable. The total number of random parameters is therefore $n_p (n_p =$ number of floors) for the beams and $n_p + 1$ for the columns, i.e. $2n_p + 1$. The coefficients of variation (COV) have normally been given the values of 0·15 for the beams and 0·20 for the columns, respectively. Tests show that in some cases the resisting moments in the beams are not predictable with any great accuracy, especially for the negative moment, due to the variable participation of the upper slab reinforcement. To explore the sensitivity of results to this source of uncertainty, a test case has been run, with all other parameters kept constant except for the COV of the beams, increased from 0·15 to 0·25.

No data have been found in the literature concerning the possible correlation structure among the strength parameters across a reinforced concrete frame. A reference structure has therefore been assumed on subjective bases, considered to be representative of a mild degree of correlation. According to this assumption, the correlation values have been set to 0·5 between the strengths of the beams and of the columns of the same floor, and to 0·25 between the strengths of the beams and of the columns of any two floors. Since the relevance of the amount of correlation between the resistances in the particular structures examined is not obvious, the extreme assumptions of $\rho = 0$ and $\rho = 1$ between all random variables have also been adopted for one of the examined cases.

3.4.4. *Analysis of the results*

The results are given in Figs 3.5 to 3.15 in terms of the fractile values of the peak ductility demands in the plastic hinges for the different frames and as functions of the design parameters q and γ_{Rd}. These two parameters have been assigned the values of $q = 3{\cdot}75$ and $q = 3{\cdot}00$, which are given in EC8 for ductility class M structures, regular and irregular along the height, respectively; and $\gamma_{Rd} = 1{\cdot}20$ and 1·00. The first value corresponds to ductility class M structures, the second one has been used to assess the effect of the absence of capacity design. The values plotted in the figures are obtained starting from equations (3.18) and (3.14), for the two cases of non-zero–mean and zero–mean response, which correspond to considering or disregarding the presence of the vertical loads, respectively. The curvature ductilities are calculated from the fractile peak values of the plastic rotations θ_{kk} by means of the expression

$$\mu_\phi = 1 + \frac{\theta_{kk}}{l_p \phi_y}$$

in which l_p is the plastic hinge length and ϕ_y is the curvature at yield. Each figure refers to two designs, made for $\gamma_{Rd} = 1{\cdot}00$ and $\gamma_{Rd} = 1{\cdot}20$; the upper diagrams give the ductility demands in the columns, the lower ones those in the beams. Unbroken lines indicate randomness in the strengths being

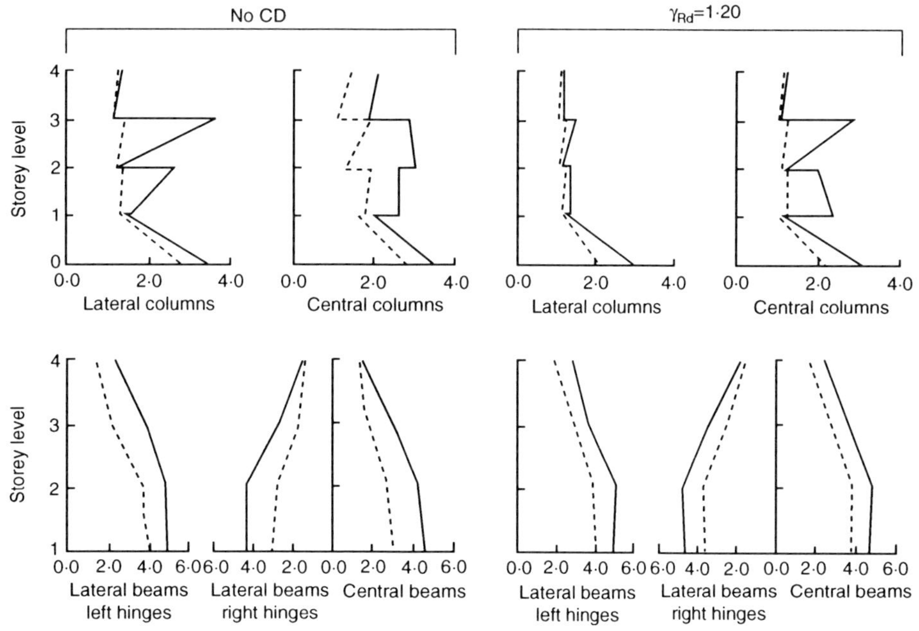

Fig. 3.5. Maximum ductility demand in columns and beams for 4-storey regular frame (a): q = 3·75, gravitational loads are considered (dotted lines = deterministic structure; continuous lines = stochastic structure); COVs of resistances and correlation structure: reference case

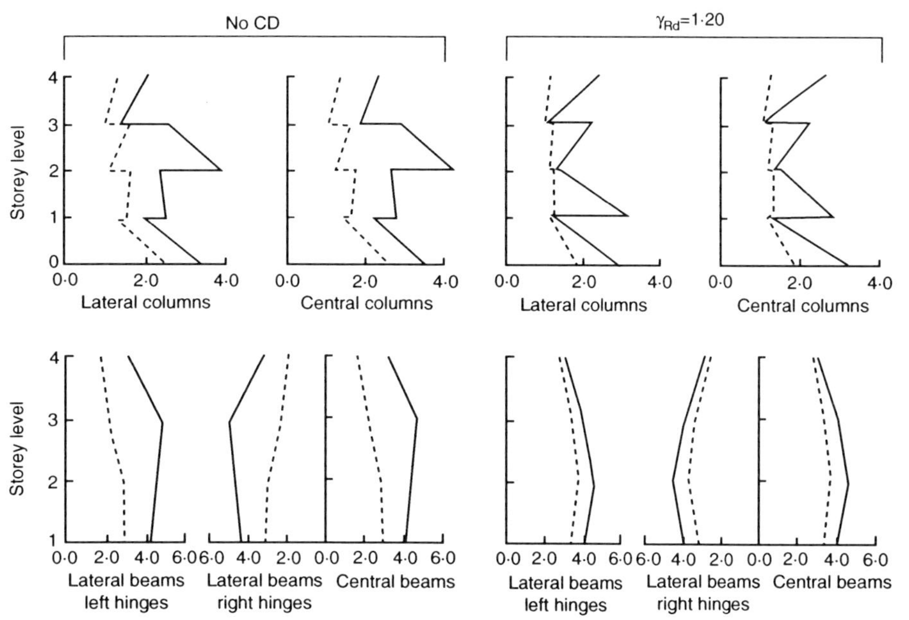

Fig. 3.6. Maximum ductility demand in columns and beams for 4-storey regular frame (a): q = 3·75, gravitational loads are ignored (dotted lines = deterministic structure; continuous lines = stochastic structure); COVs of resistances and correlation structure: reference case

considered, dotted lines deterministic structures. As expected, in all cases and in every section consideration of the randomness of the strength leads to an increase in the peak ductility demand.

Figs 3.5 and 3.6 refer to the 4-storey frame, with (Fig. 3.5) and without (Fig. 3.6) the presence of the vertical loads. The comparison shows that the differences between the two cases are rather small, which implies that for the purpose of calibrating the safety elements the vertical loads could be ignored, at least for moderate values of them. If this result could be generalized, the simplification of the analysis would be considerable. The important point from Fig. 3.5 is that a γ_{Rd} = 1·20 is not enough to protect columns from hinging, when the randomness in the strength of the elements is taken into account. The situation is obviously much worse if columns are designed for the action effects deriving from the analysis, i.e. with no consideration of CD (left-hand side of Figs 3.5 and 3.6).

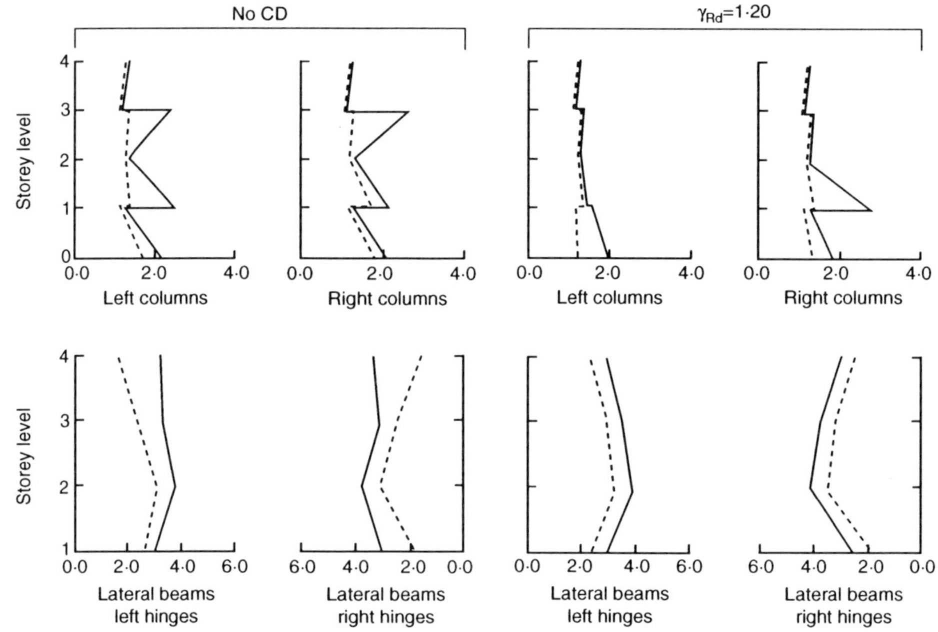

Fig. 3.7. Maximum ductility demand in columns and beams for 4-storey irregular frame (b): $q = 3{\cdot}00$, gravitational loads are considered (dotted lines = deterministic structure; continuous lines = stochastic structure); COVs of resistances and correlation structure: reference case

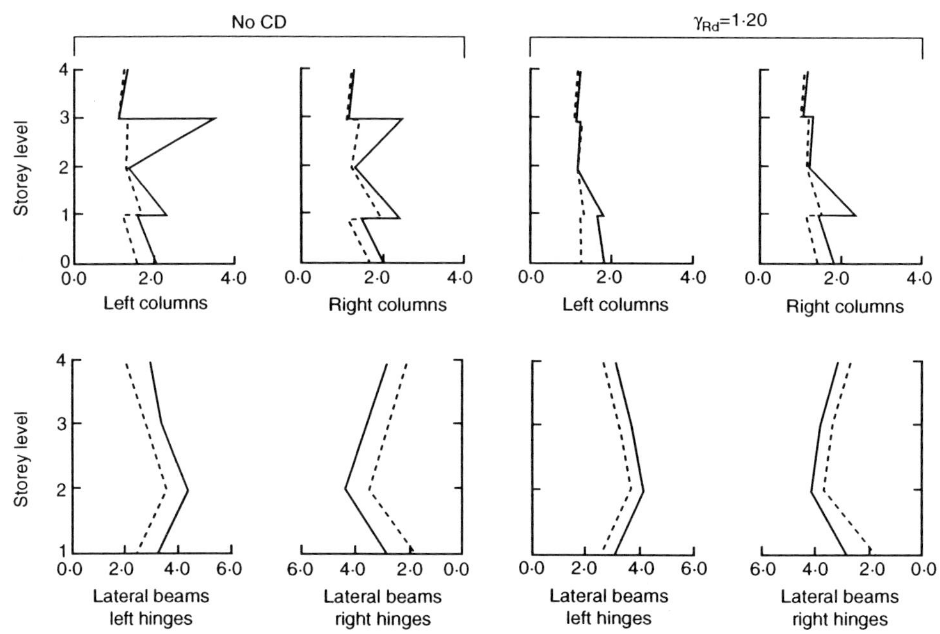

Fig. 3.8. Maximum ductility demand in columns and beams for 4-storey irregular frame (b): $q = 3{\cdot}75$, gravitational loads are considered (dotted lines = deterministic structure; continuous lines = stochastic structure); COVs of resistances and correlation structure: reference case

Figures 3.7 and 3.8 give an unexpected result: the so-called irregular (according to EC8 and other major codes) 4-storey frame behaves better, for the same q, (that is, for the same global design force) than the corresponding regular one, and the response is not noticeably affected by the 25% increase of the forces due to the change from $q = 3{\cdot}75$ to $q = 3{\cdot}00$.

The results for the 8-storey frame in Fig. 3.9 are more favourable than for the corresponding case of the 4-storey (Fig. 3.5) and show that a 'safety element' $\gamma_{Rd} = 1{\cdot}20$ is sufficient to cover the effect of the randomness and to provide almost complete protection from yielding, with the exception of the base section, which is particularly sensitive to the randomness of the strength. Not shown are the results for the same frame designed and checked without consideration of the vertical loads: in this case also (as for the 4-storey frame) the differences in the ductility demand are practically

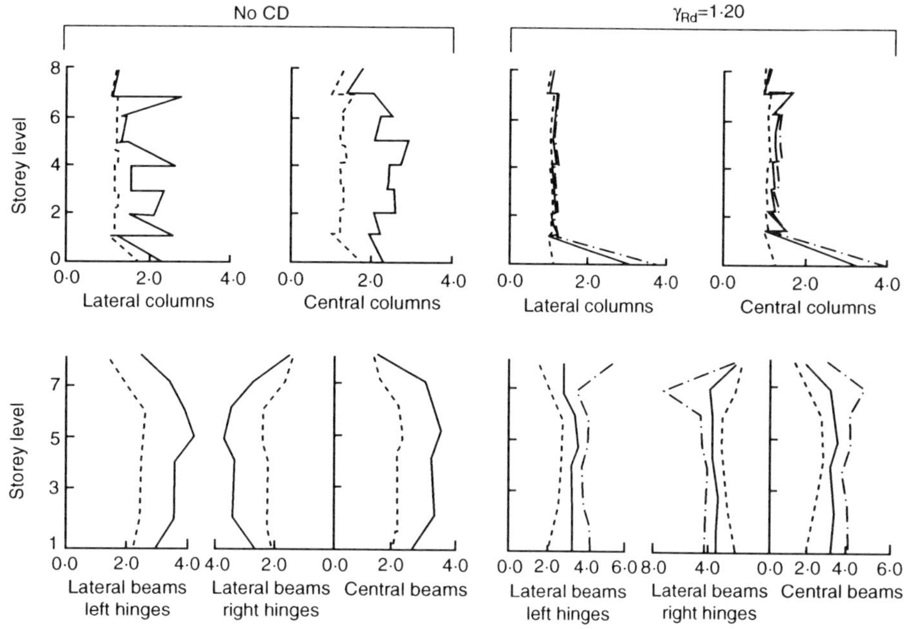

Fig. 3.9. Maximum ductility demand in columns and beams for 8-storey regular frame (c): $q = 3{\cdot}75$, gravitational loads are considered (dotted lines = deterministic structure; continuous lines = stochastic structure — reference case, dash–dot lines = stochastic structure — $\sigma_{beams} = 0{\cdot}25$); Correlation structure: reference case

negligible. In the same figure the results for the COV of the resistance of all the beams increased to 0·25 are also reported (for the case of $\gamma_{Rd} = 1{\cdot}20$). It is noted that the larger variability in the strength of the beams does not appreciably affect the ductility demand in the columns, which remain protected from yielding by the γ_{Rd} factor adopted. The influence on the beams is quite noticeable, as expected, and the increase of ductility demand for σ rising from 0·15 to 0·25 is of the same order as the increase for σ rising from zero to 0·15.

The effects of the 'irregularity' in elevation for the 8-storey frame with two bays missing, are shown in Figs 3.10 and 3.11. The conclusions duplicate exactly those found for the 4-storey frame: if the particular form of irregularity is taken into account at the design stage, the actual behaviour is no worse (or even a little better) than that of a geometrically regular structure. The value of $\gamma_{Rd} = 1{\cdot}20$ appears to be adequate to protect the columns, and little sensitivity is shown for a variation of 25% of the design seismic action.

Figure 3.12 still refers to the 8-storey regular frame, and shows the effect of different amounts of correlation between the flexural resistance of all the elements. It is not particularly significant for the columns, while for the beams the difference between the two extreme cases of $\rho = 0$ and $\rho = 1$ is quite noticeable. For both beams and columns, at all locations, the total lack of correlation represents the most unfavourable situation, and the values assumed for the reference case give results which are close to those for $\rho = 0$.

A case of large permanent loads is illustrated in Figs 3.13 and 3.14 which refer to the 3-storey frame described in Fig. 3.2. The results for the case in which the permanent loads are taken into account according to the procedure contained in EC8 are given in Fig. 3.13. It is seen that the CD procedure associated with a $\gamma_{Rd} = 1{\cdot}20$ is not capable of eliminating yielding in the columns, which is fairly uniform along the height and corresponds to a ductility of about 2 (with the usual exception of the base section). Also shown in Fig. 3.13 are the results for σ_{beams} increased from 0·15 to 0·25. As already observed for the 8-storey frame (Fig. 3.9), this increase is almost irrelevant for the columns, while it produces noticeable increases of ductility demand in the beams. The results for the same frame,

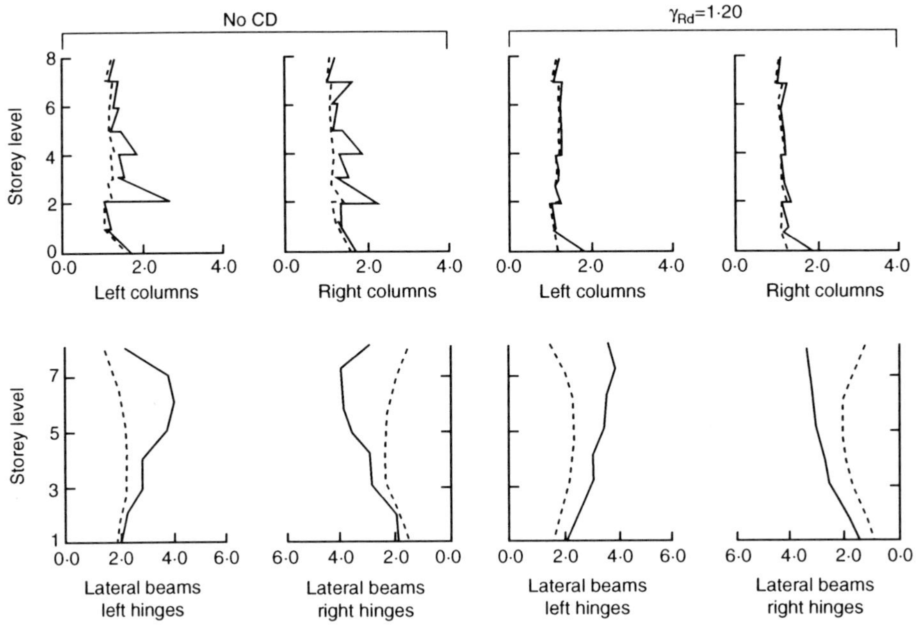

Fig. 3.10. Maximum ductility demand in columns and beams for 8-storey irregular frame (d): q = 3·00, gravitational loads are considered (dotted lines = deterministic structure; continuous lines = stochastic structure); COVs of resistances and correlation structure: reference case

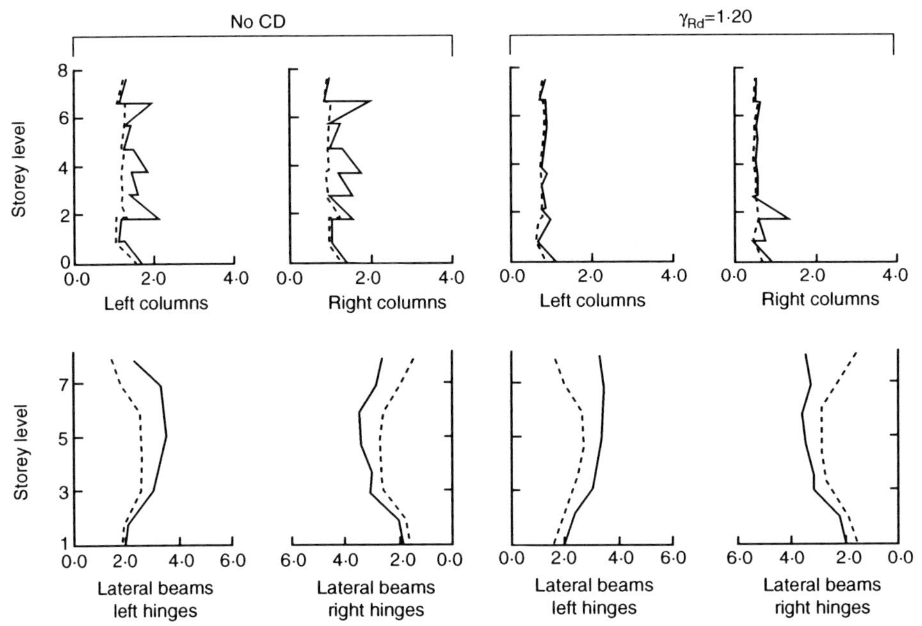

Fig. 3.11. Maximum ductility demand in columns and beams for 8-storey irregular frame (d): q = 3·75, gravitational loads are considered (dotted lines = deterministic structure; continuous lines = stochastic structure); COVs of resistances and correlation structure: reference case

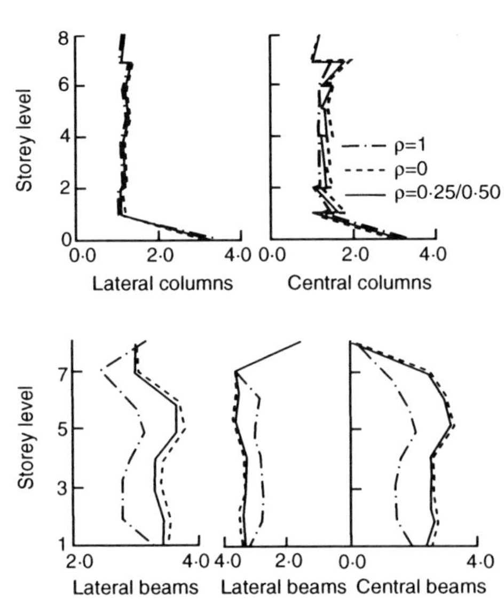

Fig. 3.12. Maximum ductility demand in columns and beams for 8-storey regular frame (c): q = 3·75, gravitational loads are considered; analysis of the effect of different correlations between resistances; COVs of resistances: reference values

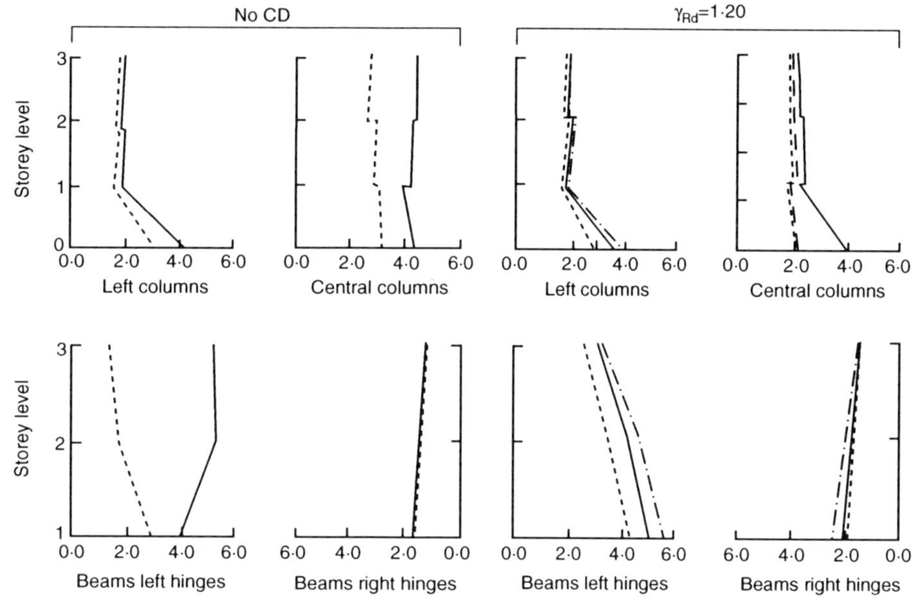

Fig. 3.13. Maximum ductility demand in columns and beams for 3-storey frame (e): q = 3·75, gravitational loads are considered (dotted lines = deterministic structure; continuous lines = stochastic structure — reference case, dash–dot lines = stochastic structure — σ_{beams} = 0·25); Correlation structure: reference case

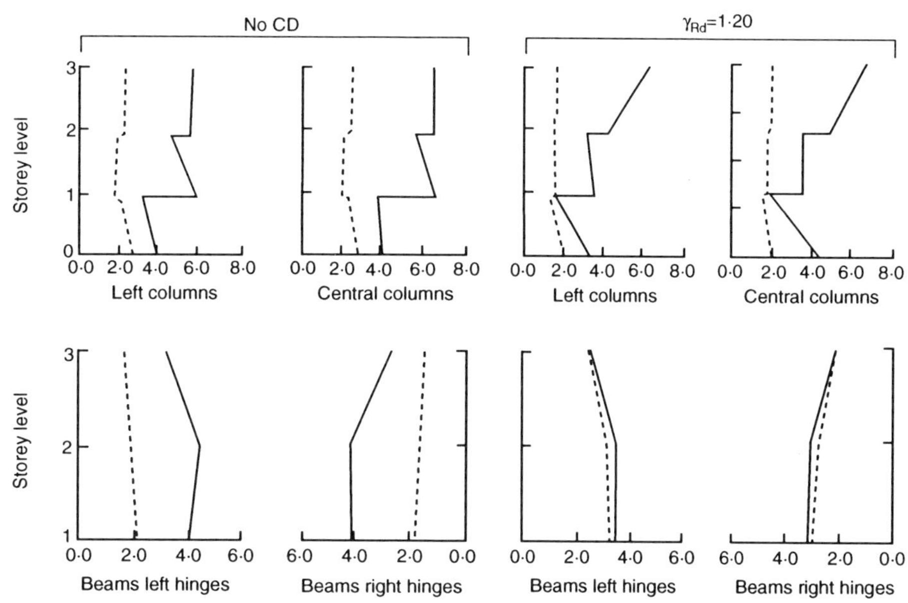

Fig. 3.14. Maximum ductility demand in columns and beams for 3-storey frame (e): q = 3·75, gravitational loads are ignored (dotted lines = deterministic structure; continuous lines = stochastic structure)

but designed and checked in the absence of the vertical loads, are presented in Fig. 3.14. It appears from the figure that if the randomnesses were ignored, the differences in the ductility demands for the cases of no CD and γ_{Rd} = 1·20 would be small with respect to the previous case. The introduction of randomness, however, drastically changes the picture with regard to the columns, leading to the conclusion that in cases of large permanent loads, calibration studies need to incorporate their effect.

Finally, Fig. 3.15 illustrates the effects of varying the degree of correlation between the element resistances. As already observed for the case of the 8-storey frame, the worst situation occurs for $\rho = 0$. Overall, the influence of the correlation is seen to be modest, except in those cases (columns of the frame without vertical loads) where the randomness effects are particularly significant by themselves.

3.5. Conclusions

The applications presented in this chapter confirm the effectiveness of the

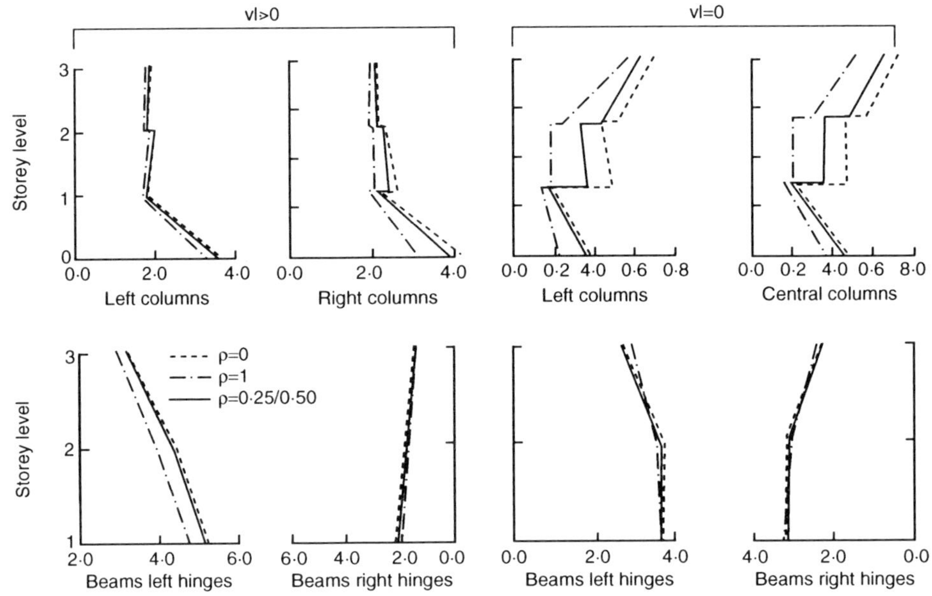

Fig. 3.15. Maximum ductility demand in columns and beams for 3-storey frame (e): q = 3·75, with and without gravitational loads; analysis of the effect of different correlations between resistances; COVs of resistances: reference values

calibration studies of the safety elements used in seismic design to ensure a satisfactory system behaviour.

All aspects of a random nature which the capacity design approach covers by means of empirically established 'safety coefficients' can be treated rigorously in the context of a reliability system approach; these include the randomness inherent in the input, described as a stochastic process, and the randomness of the various mechanical parameters of the structure, including possible correlation existing among them. Investigation of the latter two effects by means of alternative reliability methods, such as for example, the category of simulation techniques, would be prohibitive from the point of view of computational demand.

The example structures have been chosen essentially to demonstrate the flexibility of the ESL method to include specific aspects, like the presence of comparatively large gravitational loads, and a simple form of irregularity along the height. No attempt has therefore been made to arrive at values of CD factors to be directly proposed for code use. On a qualitative ground however, the results obtained for the examined cases confirm the order of magnitude of the CD factors obtained in other calibration studies.

Not unexpectedly, the randomness of the resistance properties has been found to have significant effects on the ductility demand of both columns and beams: the effects become more pronounced the larger the relative importance between the seismic and the vertical loads and, obviously, when the randomness is added to a situation of already insufficient protection of the columns. When columns are sufficiently protected by the use of an adequate 'overstrength factor', increasing the variability of beam resistances has little effect on the ductility demand on them, while the demand on the beams themselves increases significantly. It may be worth underlining the fact that the effect just discussed refers to the variability of beam strengths about their average value (which is assumed to be deterministically known) and has therefore to be distinguished from the deterministic physical fact that negative beam end moments are actually larger than the design ones due to the contribution of the slab reinforcement.

The correlation existing among the mechanical properties of the elements of which a reinforced concrete frame is composed has been little investigated thus far, and no data have been found in the literature. The effects of different correlation patterns can be easily introduced in the ESL procedure adopted, however, and a sample exploration has been made to gain first indications on the relevance of this aspect. Three patterns have been considered: full ($\rho = 1$), zero ($\rho = 0$) and an intermediate one, based on judgement. The results obtained for the cases examined indicate (again, not unexpectedly) that the most unfavourable situation occurs for $\rho = 0$, but the effect of varying ρ is not particularly significant, except in those cases in which the randomness effects by themselves are more pronounced (higher variabilities of member strengths, higher variability in the action).

4. Measures of seismic performance

4.1. Basic concepts and scope

Most of the modern seismic codes specify two fundamental performance criteria for RC structures:

- no collapse and/or excessive damage (under the design earthquake)
- limitation of damage (under an earthquake with higher probability of occurrence than the design one).

It is worth pointing out that specific criteria for the implementation of the above principles vary from code to code, and inconsistencies are not uncommon. However, the terms *collapse* (or *failure*) and *damage* are more or less common to all codes and some correlation with the ultimate and the serviceability limit states is usually made. Hence, if a quantification of seismic performance criteria is sought, it is necessary to express damage in a quantitative form, with failure corresponding to the maximum degree of damage which a structure can sustain. The present chapter focuses on concepts and procedures for evaluating damage indices for reinforced concrete structures, with a view to assessing their seismic performance.

Although analytical and experimental research into the seismic behaviour of concrete structures has flourished during the past 40 years, systematic attempts to estimate quantitatively the degree of seismic damage that a structure has suffered have appeared only during the past 15 years or so. Of course, the use of the well-known *ductility factors* as *damage variables* was suggested a long time ago (Blume *et al.*, 1961). However, the incorporation of damage variables into actual *damage indices* and, even more, the attempt to calibrate these indices against available experimental data have been carried out only during the past 15 years.

Before proceeding to the clarification of the terms 'damage index' and 'damage variable', it is worth mentioning some typical situations, where the use of some sort of damage indicator is warranted. Such a typical situation is post-earthquake damage assessment, which normally consists of two stages: a rapid initial stage (visual screening), during which the main goal is to decide whether a building is habitable or not, and a subsequent, more detailed stage (structural assessment), during which the required measures for repair and/or strengthening have to be defined. Other situations where a theoretical estimate of the degree of damage is required are reliability studies of existing structures and earthquake damage scenarios, based on which a decision can be made as to whether or not a structure should be strengthened (pre-earthquake strengthening), and also seismic performance predictions for novel types of structure, especially those of high importance. Such predictions may serve as an invaluable aid in the seismic design of these structures. It is understood that the degree of sophistication warranted in the evaluation of a certain damage index always depends on the situation in which it is used.

4.2. The notion of damage parameter and damage index

4.2.1. Identification of the degree of damage in RC members

Damage in reinforced concrete (RC) members is generally related to cracking and, at a subsequent stage, to crushing, first of the cover ('spalling') and later of the confined core, which corresponds to failure of the member. Following spalling of the cover concrete, other failure modes may precede crushing of the confined core, for instance buckling of and possibly fracture of longitudinal bars, or loss of anchorage (bond failure).

Ultimate conditions in RC members are not easy to define, even under predominantly flexural conditions. A number of criteria have been used.

(*a*) A more or less arbitrary strength drop (values ranging from 10–30% have been proposed), referring to the load P — deflection δ or the moment M — rotation θ curve. This approach is clearly subjective and perhaps inappropriate.
(*b*) Failure of confinement, corresponding to fracture of at least one hoop or spiral. Recent tests (Priestley, 1993) have shown that fracture of transverse reinforcement does not represent failure but the onset of cyclic strength degradation, at a rate which probably depends on axial load, size and arrangement of longitudinal reinforcement, and member proportions.
(*c*) Attainment of an ultimate tensile strain ϵ_{su} in longitudinal reinforcement. This is rarely the case in North America or New Zealand, but may be more so in Europe, especially when A or B steels (as specified in the Eurocodes) are used.
(*d*) Onset of buckling of longitudinal reinforcement. This is followed within a few cycles by fracture of longitudinal reinforcement and rapid strength degradation.

4.2.2. *Damage parameters and criteria for their selection*

In general, damage in RC is related to irrecoverable (inelastic) deformations, therefore any damage variable (damage parameter) should preferably refer to a certain deformation quantity. Such quantities are strain (compressive/tensile) and curvature, but displacement quantities are equally good candidates, namely rotations (at member ends), horizontal storey displacements (vertical members in RC buildings are typically assumed to deflect by equal amounts at each floor level), and relative displacements between adjacent storeys (inter-storey drifts). The last two quantities are, as a rule, used as global damage variables, while curvatures or rotations are used for the characterization of local damage. Although related in a less straightforward way to damage, forces are also sometimes used as damage variables (base shears, storey shears, member resistances). Last but not least, the energy absorbed or the energy dissipated during inelastic reversed cyclic loading of a RC member (or structure) is another meaningful damage variable.

All the damage variables mentioned in the previous paragraph involve a certain structural quantity. However, when referring to damage, it is often meaningful from the practical point of view to express it in monetary terms, typically as the cost required to restore a member or a structure to its initial (pre-earthquake) state. Such economic damage variables are indispensable in earthquake insurance problems and quite useful in many other situations involving seismic risk assessment. It is understood that the values of economic damage variables are closely related to those of the structural ones, nevertheless the establishment of such a relationship is a topic still open for research, as will be discussed in more detail in later sections.

4.2.3. *Definition of the damage index*

The terms ‘damage variable’ and ‘damage index (indicator)’ are usually interchangeable in the literature, with the possible exception of the past five-year period. Strictly speaking and in order to avoid difficulties of interpretation, a damage index (D) is a quantity with zero value when no

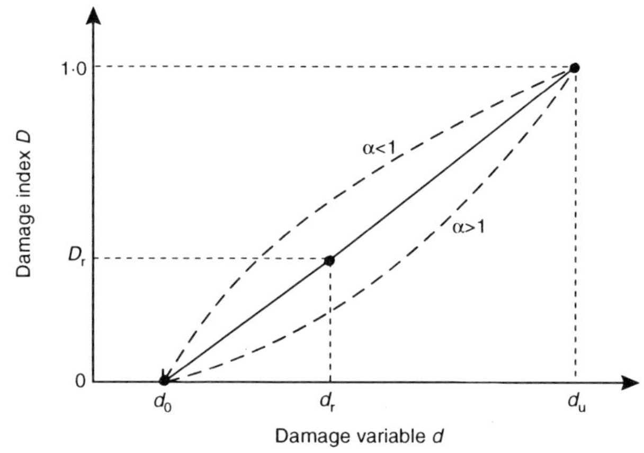

Fig. 4.1 Relationship between damage variable and damage index

damage occurs and a value of 1 (or 100%) when failure or collapse occurs. Furthermore, a damage index may (and in fact should, as will be shown later) involve more than one damage variable.

The relationship between a damage variable (d) and a damage index (D) is illustrated in Fig. 4.1, for the simple case where D is dependent on a single variable d. As mentioned previously, d can be a deformation, displacement, force or energy quantity. In fact any of these quantities may be introduced in a non-dimensional form, for instance curvature or rotation may be normalized to their yield values, therefore becoming ductility factors. As shown in Fig. 4.1, for $D = 0$ it is $d = d_0$ and typically $d_0 > 0$, which implies that there is a threshold value for the damage variable, below which virtually elastic behaviour occurs, in the sense that no permanent deformation is visible, therefore no damage is detected.

Besides d_0 there are at least two more values of d that may be of particular importance, especially from the engineering point of view. One is d_u, the value of the damage variable for which collapse is assumed to occur ($D = 1$). There is significant difficulty, as well as space for subjective judgement in determining appropriate values for d_u and different researchers have used different approaches, usually on the conservative (or even overly conservative) side. Equally important for the engineer who must make the appropriate decisions, is the value d_r, beyond which some repair of the member (or the structure) is required, to restore it to its initial condition. Clearly, the selection of the appropriate value for d_r is not an easy task, especially in the case of complex structures such as those of RC. It has to be noted that economic and occasionally even political considerations are involved in making this decision.

Further complications arise with regard to the shape of the curve $D = f(d)$, since the available experimental data are usually either insufficient or inconclusive. A reasonable, as well as physically sound choice is the function (Oliveira, 1977, Powell and Allahabadi, 1988).

$$D = (d_{cal} - d_0)^{\alpha}/(d_u - d_0)^{\alpha} \tag{4.1}$$

where d_{cal} is the value of the damage variable calculated from analysis and α is an exponent which, in the absence of conclusive experimental data, may be taken as equal to unity, although $\alpha > 1$ appears to match better the limited data available to date. It has to be pointed out here that the relationship of Fig. 4.1 is also valid for economic damage variables, in which case the economic damage index is preferably defined as the ratio of the required cost of repair to the corresponding cost of reconstruction (replacement), i.e. the cost corresponding to demolition of the damaged structure and construction of a new identical one.

4.3. Classification schemes and examples of damage indices

Numerous damage indices have been proposed in the literature, and within the framework of the present report it is appropriate to review them, classified in the most meaningful way possible. In fact, there are quite a few different ways to classify these indices, some of them already used in previous similar reports (see Chung *et al.*, 1987). Detailed discussions of the damage indices proposed in the literature can be found in recent state-of-the-art reports (Chung *et al.*, 1987, Ang, 1989, CEB TG III/6, 1994, Williams and Sexsmith, 1995), while a comprehensive list of references including studies on seismic damage indices for RC structures is given at the end of this chapter. The focus of the presentation which follows is on clarifying the different approaches that can be used for defining damage indices in terms of damage parameters, and on possible advantages and disadvantages of each approach. Several examples of indices are given in each case, but no explicit presentation of individual indices is made.

4.3.1. Local and global indices

If the reference point is the part of the structure whose damage is described, the different categories are local, intermediate and global indices, depending on whether they refer to a single member, to a subassemblage of the structure (a storey in a building, or more generally a substructure), or to the structure as a whole. It is easily understood that the determination of the damage index becomes less accurate as one shifts from a critical region to the structure in its entirety. Furthermore, it has to be pointed out that some damage indices (for instance displacement ductilities) can be used both as local and global indices, while others (for instance inter-storey drifts or curvature ductilities) can only be used as either global or local.

Local indices may involve a single damage parameter, such as maximum deformation (curvature or rotation) or dissipated energy, or two or more parameters. For example, Banon and Veneziano (1981) have used the flexural damage ratio (flexural stiffness divided by the reduced secant stiffness at maximum displacement) in combination with the dissipated energy; Hwang (1982) has used the previous two parameters and the ductility ratio; Park and Ang (1985) have used rotational ductility and dissipated energy; Fardis (1995) suggested a modified form of the Park–Ang index, where the maximum rotation is substituted by the peak value of the member deformation energy; Mizuhata and Maeda (1989; first published in Japanese in 1983) have combined the number of load cycles with the corresponding displacement ductility; Chung *et al.* (1987–1989) have also used the number of load cycles together with a damage-based hysteresis model; Bracci *et al.* (1989) have suggested a damage index equal to the ratio of 'damage consumption' (loss in damage capacity) to 'damage potential' (capacity), defined as appropriate areas under the monotonic and the low-cycle fatigue envelopes.

Global indices can be defined in terms of a global parameter, for instance global ductility factors (based on storey displacements), such as the one based on roof displacement (Roufaiel and Meyer, 1987), or softening indices relating the initial fundamental period of the structure to the final one (DiPasquale *et al.*, 1987–1990); the latter approach can be used with two (Mork, 1992) or more modes, in order to detect concentration of damage in the top or the bottom of the structure. Intermediate and global indices can be defined as weighted averages of individual member indices (taken at each storey of a building or for the entire structure). The weighting factors may involve the energy dissipated by a member (Reinhorn *et al.*, 1988, Seidel *et al.*, 1989), or the tributary gravity load of a member (Bracci *et al.*, 1989); both approaches generally tend to give more weight to the members of the lower storeys, which is conceptually correct, but fail to recognize that failure of a (soft) storey typically means failure of the structure.

4.3.2. Deterministic and probabilistic indices

Depending on the mathematical approach used to determine the damage index, one may distinguish between deterministic and probabilistic indices. Given the uncertainties involved both in the seismic action and in the mechanisms of RC members' resistance to reversed cyclic loading applied at high strain rates, it appears that the probabilistic approach might be the most appropriate choice. However, most of the effort so far has concentrated on deterministic damage indices, mainly because of their relative simplicity, of the possibility of being used directly in a practical context, and, of course, because the computational cost required for their determination is significantly lower than that required for the probabilistic indices.

Most of the damage indices used in the references given at the end of the chapter are deterministically derived. Probabilistic studies involving damage indices have been presented by Banon and Veneziano (1982), Ciampoli *et al.* (1989), DiPasquale and Cakmak (1990), Jeong and Iwan (1988), Nielsen *et al.* (1992), Mork (1992), and others.

4.3.3. Classification based on type of analysis

The majority of damage variables proposed in the literature require some sort of analysis for their determination, although this is not one of their indispensable features (other approaches based on inspection and in-situ measurements have also been suggested). Therefore, another categorization scheme can be based on the type of analysis required, that is linear-elastic or inelastic, or no analysis at all. Furthermore, one might distinguish between static or dynamic analysis in each of the first two cases. Ideally, inelastic dynamic (time-history) analysis should be used in estimating seismic damage which, by definition, refers to members having entered the post-elastic range of their response being subjected to a time-varying action such as the earthquake. However, both the expertise and the cost required to efficiently carry out such an analysis for a multi-storey RC structure other than a well-designed and detailed, flexure-dominated plane frame is a very difficult task, even for research purposes. Therefore, other, less demanding types of analysis should not be ruled out, especially whenever damage estimation studies are undertaken within the design office environment. Finally, it has to be clarified that 'no analysis at all' means that the damage index is estimated empirically, typically using simple formulae correlating earthquake measures, such as magnitude or intensity, to damage to different types of structures. This empirical approach to vulnerability studies is quite useful practically, even more so in the case of structures for which analysis is cumbersome or even not feasible. For this reason it has been quite popular in the case of load-bearing masonry structures made of various materials, but it has also been used for RC structures, mostly buildings.

The problems involved in the analytical estimation of damage indices, such as model selection and failure criteria, are further discussed in section 4.4.

4.3.4 Structural and economic indices

As already mentioned in the previous section, it is quite interesting to express damage in monetary terms, by means of an economic damage variable. Therefore, another categorization scheme for damage indices can be based on whether they are structural or economic. The corresponding variables for each category have already been identified in the previous section. Here it has to be pointed out that most of the research so far has concentrated on structural damage indices, while in the limited number of studies involving economic indices, an almost exclusively empirical approach has been used. The typical form of the economic damage index is the ratio of the cost required for the repair of the member or the structure to the corresponding cost of reconstruction (replacement). It has to be emphasized that by following this approach, it is possible that the value of the economic index exceeds unity, while the structure under consideration has not physically collapsed. In fact, depending on the assumptions made, the values of economic and structural damage indices estimated for a specific structure can be different.

The economic damage index is typically used for earthquake insurance decision making, as it provides a convenient way of defining the appropriate premium. A discussion of loss assessment methods based on appropriate selection of hazard and vulnerability models can be found in the recent report by Dolce *et al.* (1995). Among the few studies involving correlation of an analytically calculated structural index and a corresponding economic damage index, are those by Kappos *et al.* (1991, 1996) where the rotational ductility factor at a critical section is correlated to the required cost of repair (considering different repair techniques) for RC members, and that by Gunturi and Shah (1992) where the aforementioned Park–Ang (1985) index is related to the replacement cost of RC members.

4.3.5. Classification based on the approach used in defining the damage index

The numerous structural damage indices proposed in the literature are typically based on either of the following approaches.

(*a*) The demand imposed by the earthquake with respect to a certain structural quantity (for instance deformation or energy) is related to the corresponding capacity of the member or the structure as a whole (supply/demand approach).

(*b*) The degradation of a certain seismic variable (strength, stiffness energy dissipation) is compared with a predetermined critical value, usually expressed as a percentage of the initial value corresponding to the undamaged state.

These different approaches to deriving seismic damage indices can also serve as a basis for a further classification scheme.

The previously mentioned index by Bracci *et al.* (1989) is a typical example of the capacity/demand approach; other examples involve the cumulative displacement index suggested by Bertero and Bresler (1977), and the combined indices of Mizuhata and Maeda (1989) and Chung *et al.* (1989) briefly discussed in section 4.3.1. The main difficulty in defining these indices is the estimation of 'capacity', especially when cumulative damage parameters are involved (such as energy or cumulative deformation). A common, though crude, approximation is to define capacity with respect to monotonic loading, for instance as the area under the monotonic load–deformation curve. Some of the aforementioned models involve the parameter 'number of cycles to failure' which is very difficult to obtain since data from low-cycle high-amplitude fatigue tests (which are the only ones pertinent to seismic conditions) are limited and the scatter is large. This appears to be the reason that indices such as those by Chung *et al.* (1989) and Bracci *et al.* (1989), although conceptually very attractive, have not been extensively used.

Although several damage parameters based on degradation of a certain seismic variable have been proposed, involving stiffness (Lybas and Sozen, 1977, Banon *et al.*, 1981, Toussi *et al.*, 1984, Roufaiel and Meyer, 1987), or period of vibration (DiPasquale *et al.*, (1987–1990, Mork, 1992), their expression in the form of a damage index ($0 \leq D \leq 1$) requires the definition of the limiting value of degradation after which failure occurs, which is typically at least as difficult to define as the deformation or energy dissipation capacity. Therefore it may be stated that this category of indices is primarily used in project-specific system identification studies, particularly those involving instrumented buildings.

4.3.6. Structural and non-structural elements

Although the focus of the present document is on RC structural members, it should not be forgotten that the cost of these members hardly ever exceeds one-quarter of the total cost of a typical building. It is therefore important to establish reliable damage indices for non-structural elements, such as cladding members, partition walls and the various installations, not to mention the content of certain buildings which might by far exceed the cost of the structure itself. Among the usual 'non-structural' members, the most vulnerable ones are those which can not easily follow the deformation pattern of the RC structural system during the earthquake excitation. A typical member of this category is the masonry infill panel, widely used in many parts of the world, in the usual case where it is constructed in contact with the surrounding frame. The cost of repairing infill panels after an earthquake often exceeds that for RC members, therefore it is essential that reliable seismic damage indices are developed for at least this type of element.

Kappos *et al.* (1991, 1996) have suggested correlating the inter-storey drift to the required cost of repair (considering different repair techniques) for masonry infills in RC frames, while Gunturi and Shah (1992) derived loss curves for non-structural elements and the building contents as functions of the inter-storey drift and the storey acceleration, respectively.

4.4. Procedures for the determination of damage parameters

4.4.1. Level of discretization used in the analysis

The level of damage definition is related to the level of discretization in the analytical model, and it cannot be finer than the latter.

(*a*) Macroscopic models, such as the equivalent SDOF systems, e.g. the Saiidi and Sozen (1979) *Q*-model, and the shear beam models or shear buildings (e.g. Chopra and Kan, 1973). Only global response quantities can be calculated using these models, such as top displacements (maximum and/or cumulative values), and base shears (also storey shears and displacements in the second category). These models are possibly appropriate for rough estimates of global damage.

(*b*) Member-type models (element-to-element discretization), wherein the finite element typically coincides with the actual structural member. These include

- lumped plasticity models, e.g. the two-component model by Clough *et al.* (1965), and the one-component model by Giberson (1967)
- distributed plasticity models, e.g. the plastification zones model by Soleimani *et al.* (1979), and the distributed flexibility models by Takizawa (1976) and by Park *et al.* (1987).

Member response quantities, such as deformations, plastic hinge rotations and/or curvatures, as well as the corresponding cumulative quantities may be calculated, in addition to global quantities. These models are potentially adequate for member damage estimation, provided appropriate constitutive laws are used for the plastic hinge regions.

(*c*) Microscopic models, including among others

- fibre or layer or filament models, at the cross-section level, or at both the cross-section level and along the longitudinal axis (multi-slice fibre models), first used in inelastic dynamic analysis of RC structures by Mark (1976)
- finite element (FE) models, such as the plane stress element for cyclic shear proposed by Stevens *et al.* (1991).

Stresses and strains, as well as curvatures at various stations may be monitored and used for estimating local damage or member damage (by integration). Estimation of member and especially of storey damage is usually inconvenient from the computational point of view in FE analysis (substructuring techniques may alleviate part of the problem).

A detailed critical review of the foregoing types of model, with the exception of those based on the (continuum) FE approach, which is generally not suitable for studying the inelastic dynamic response of large-scale structures such as multi-storey RC buildings, can be found in the recent report by the CEB Task Group III/6 (CEB 1994). It is pointed out that most of the foregoing models are best suited for line (1D) elements such as beams and columns, and less reliable when applied to essentially planar systems such as structural walls; their reliability generally decreases with the aspect ratio of the member to be modelled.

4.4.2. Coupling between hysteresis law and degree of damage

The 'ideal' analytical model accounts both for the effect of the degree of damage (expressed through an appropriate index *D*) on the non-linear constitutive law, and for the interrelation between evolution of damage and inelastic response quantities (the ones selected as damage variables, *d*).

A simple, yet conceptually attractive in the light of the foregoing remarks, model is that proposed by Wang and Shah (1987), which uses a damage index that is a function of the cumulative cyclic inelastic deformation for describing strength degradation with cycling.

4.4.3. Failure criteria used in the analysis

4.4.3.1. Local failure. As discussed in section 4.2.1, at member level, buckling of longitudinal steel appears to be a good candidate as a local failure criterion, at least for flexure-dominated members. However, no generally accepted analytical method for estimating the onset of buckling appears to be available. An interesting, though relatively limited in scope, procedure has been suggested by Papia and Russo (1989), who have expressed the ultimate concrete strain at the onset of buckling as a function of all the relevant parameters, hence the ultimate curvature (ϕ_u) can be easily calculated. Alternatively, a model for reinforcing bars accounting for inelastic buckling can be used (Monti and Nuti, 1992), in which case failure has to be defined on the basis of a pre-specified percentage of section strength drop. Other approaches for defining failure at section level include selecting (more or less arbitrarily) a percentage of strength drop along the descending branch of the confined concrete stress–strain curve, or using empirical equations for the strain corresponding to first hoop fracture; a detailed discussion of the various approaches may be found in Penelis and Kappos (1996).

Having established ultimate conditions at critical section level, it is then customary to evaluate the rotational capacity of a member by introducing the equivalent plastic hinge length (l_p) concept, which presupposes availability of a reliable database of test data pertinent to the type of member studied. Several equations suggested in the literature for l_p are discussed in Penelis and Kappos (1996). A recent equation based on a database including appropriately confined columns is presented in Chapter 6, where detailed information on assessment of strength and ductility of members in existing RC structures is also given. It should be borne in mind that in addition to flexural strength and ductility, shear capacity should also be considered in assessing local failure in RC members. Unfortunately, no properly calibrated model predicting the ultimate shear deformation (γ_u) of RC beams, columns and walls is available to date; hence shear failure can only be predicted in strength terms. Priestley *et al.* (1994) have proposed a model for column shear strength under cyclic loading, wherein the 'concrete' term is a function of the cyclic displacement (or curvature) ductility; the model is also presented in Chapter 6 of the present Design Guide. Garstka *et al.* (1993) have proposed, on the basis of nine test results involving beams with three different shear spans, a flexure–shear interaction model, wherein D is expressed as a non-linear combination of damage due to shear and to flexure. Both components of D are based purely on energy absorption, using a concept previously suggested by Meyer *et al.* (1988). The former flexure–shear interaction model was based solely on energy values calculated under monotonic loading conditions.

4.4.3.2. Global failure. Local failure, however reliably predicted by analysis, does not necessarily indicate collapse of a storey or of the building as a whole. Predictions of 'failure' based on member ductility criteria can be either unconservative or conservative. Actually, if failure is defined as the state of a building, beyond which the cost of repair is greater than the cost of reconstruction, the local failure criterion may well be unconservative, in particular when the cost of repairing the 'non-structural' members dominates, which is usually the case. On the other hand, if a member which is not essential to the stability of a building (for instance a beam) fails prematurely and redistibution of actions is possible, the local failure criterion may yield conservative results. Based on the above remarks, it is considered essential to include a global failure criterion in the analytical procedure for assessing the seismic performance of RC buildings. As a rule, it is conservatively assumed that the failure of a single storey is equivalent to the overall failure of the building, although post-earthquake inspections have revealed that this is not always the case, especially in structures with a soft first storey.

Perhaps the single most important response parameter to characterize the seismic behaviour of a storey or a building is the relative inter-storey drift (Sozen, 1981), defined as

$$\Delta x_i/h_i = (x_i - x_{i-1})/h_i \tag{4.2}$$

where x_i, x_{i-1} denote the horizontal displacements of two adjacent floors and h_i the corresponding storey height. This quantity is easy to measure in tests or in actual buildings struck by earthquakes and can be correlated with available data on damage. However, it is very difficult to define a single value of drift corresponding to collapse, to apply for all buildings. Taking into consideration factors such as the vulnerability and importance of the building contents, as well as the cost of repair, it was suggested (Sozen, 1981) that an inter-storey drift of 2% might be set as the collapse limit for about three-quarters of RC buildings; this value may be included in a combined criterion of storey failure. It has to be emphasized that for this criterion to be valid, it is essential that the stiffness of the building under consideration be adequately modelled in the analysis, otherwise the procedure may become unconservative.

The inter-storey drift also serves as a measure of the effect of second-order shears and moments, which are not treated in an explicit way by most computer codes for inelastic time-history analysis, including the one used in the studies presented in the following sections. At values of inter-storey drift in excess of 2% the P–δ effect is significant, reducing the lateral force resistance and the stiffness of the vertical structural members and precipitating failure.

Another criterion for storey failure, commonly used in the cases of inelastic static analysis, as well as of limit analysis, is the formation of a 'sidesway' mechanism, involving plastic hinges at both the top and the bottom of all vertical members. The above criterion may be rather conservative in dynamic analysis because at the time a hinge forms at a certain member end, another, already yielding member may enter the unloading stage and respond with a stiffness equal to, or slightly lower than, the elastic one. Therefore, a combined criterion has been suggested by Kappos (1991a) involving both the formation of a sidesway mechanism and the occurrence of an inter-storey drift in excess of 2%. However, even in the case that a collapse mechanism does not form, a building should be assumed to have failed whenever the maximum inter-storey drift exceeds 3%, since at this stage all non-structural elements have been severely damaged and repair of the building is no longer cost-effective.

4.4.4. *Representative 'loading' history*

The effect of loading history, in particular of the influence of cycles inducing a high degree of inelasticity on subsequent cycles at lower amplitudes, has not been clarified in a definitive way. Hwang and Scribner (1984) have studied the cyclic response of beams subjected to various displacement time-histories, and concluded that strength and stiffness degradation was closely related to the maximum amplitude in each cycle, but relatively independent of the sequence in which large and small displacements were applied.

Models including the effect of loading history have been proposed by Chung *et al.* (1987, 1989), and by Wang and Shah (1987), but their calibration against experimental data appears to be rather limited.

4.4.5. *Alternative methods for damage assessment*

Although the focus of the present chapter is on the analytical estimation of damage indices, the practical use of the concept can be combined with other (non-analytical) techniques. Two such procedures are outlined below.

4.4.5.1. Use of response measurements. Takahashi *et al.* (1989) have suggested the use of crack indices for shear walls. The measured quantities included crack widths, number of cracks, and length of cracks (grid method). The corresponding damage variables were crack width, crack width ratio (elongation of member due to cracking), and crack area ratio (increase of member area due to cracking). A correlation between crack indices and deformation index (δ/δ_u) has been made based on test results.

Cracking and spalling indices for columns have been suggested by Ogawa and Shiga (1989). The measuring procedure used involved drawing on computer displays of crack patterns and spalling area boundaries. The damage variables were the equivalent cracked area ratio and the spalling area ratio which were correlated to displacement ductility.

4.4.5.2. Use of empirical vulnerability of functions. By far the most common procedure used for practical assessment of structures is through functions (or matrices) describing the damage state of a particular structural type as a function of the earthquake intensity; these functions are derived on the basis of statistical data collected after catastrophic earthquakes. The procedure can only be applied to groups of structures rather than to specific buildings, and its reliability depends on the amount and quality of the data available for a certain category of structures. The use of this procedure is almost mandatory in the case of structures which are not amenable to analysis, as are many types of masonry buildings, in particular older ones.

The current trends in vulnerability assessment are discussed in a recent report by the EAEE Group on Seismic Risk and Vulnerability (Dolce *et al.* 1995).

4.5. Evaluation of two commonly used damage indices

4.5.1. Scope of the study

As an illustration of the various concepts put forward in the previous sections, the main results will be presented from a recent comparative study by Kappos and Xenos (1996) involving the two most widely used indices, namely the ductility factor, introduced in the 1960s, and the more recent index by Park *et al.* (1985, 1987), which combines maximum deformation and hysteretic energy dissipation.

With regard to the former index the main reason for its popularity is obviously its simplicity, while the Park–Ang index's extensive use should be attributed to two main reasons (in order of importance).

(*a*) It was calibrated against a very large database; in the original version (Park and Ang, 1985) a total of 261 specimens were considered, while in a later version (Park *et al.*, 1987, 1988) the number of specimens was increased to 402 and some rather significant revisions to the main empirical coefficients were suggested.

(*b*) It is conceptually attractive, since it combines the effect of damage caused by excessive deformation (related to monotonic loading) and the effect of cyclic loading as expressed by the hysteretic energy dissipation.

It is understood that the evaluation of the combined index is computationally more demanding than that of ductility factors, and in addition it involves the determination of coefficients which can not be estimated on the basis of first principles and material constitutive laws.

In the light of the above, the main objective of the study was to assess the importance of the energy term in combined damage indices, considering structures of realistic size and hysteretic characteristics, realistic seismic inputs (base motions), and also a sufficiently rigorous analysis procedure.

4.5.2. Analytical procedure

The procedure used consisted of the following steps.

- Design of two typical ten-storey buildings, one frame and one dual (frame + wall) structure, according to the ENV Eurocode 8 (CEN, 1995), for a design acceleration A_d = 0·25g and medium ('M') ductility level (see Kappos and Antoniades, 1995).

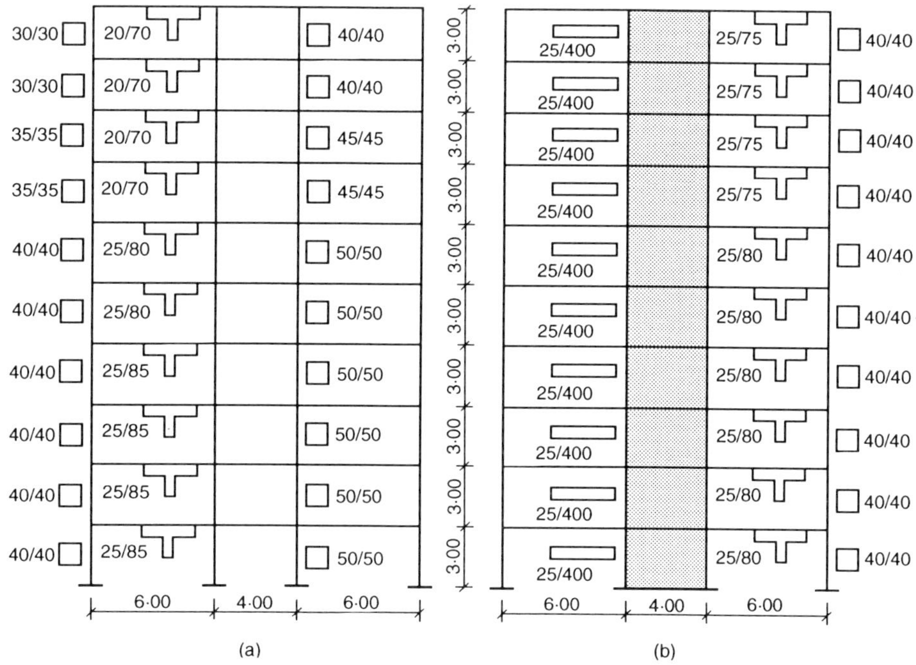

Fig. 4.2. Geometric data for the structures studied: (a) frame; (b) dual system

- Idealization of the structures as planar assemblies of line elements (Fig. 4.2), the behaviour of each element described using the distributed flexibility model incorporated in a modified version of the IDARC code (Park *et al.*, 1987).
- Selection of a suite of seven input motions (acceleration time-histories) typical for either southern Europe (Greece) or California (see Table 4.1). All motions were normalized to the intensity of the design earthquake using a modified Housner technique which has been found to reduce scatter in calculated inelastic response (Kappos, 1990).
- Analysis of the two structures for all motions, considering a duration during which 95% of the energy is released. In order to simulate the effect of a very rare seismic event producing a very long record, the effective duration of the well-known El Centro 1940 record was considered in 'duplicate', that is two identical parts of 26·5 s in sequence.
- Calculation of the required to available rotational ductility ratio at each critical region of the RC members, as well as of the corresponding Park–Ang damage index

$$D = \frac{\theta_{\max}}{\theta_{\mathrm{u}}} + \beta \frac{\int M d\theta}{M_{\mathrm{y}} \theta_{\mathrm{u}}} \tag{4.3}$$

A crucial decision with regard to the latter index is the selection of the β factor which defines the contribution of the dissipated energy to damage; as found by Park and Ang (1985), the scatter in this coefficient is very large and, for some test specimens, physically unreasonable values (less than zero, or greater than 1) are calculated. However, based on later studies by Park *et al.* (1987), it appears that a reasonable value, at least for adequately detailed members, is $\beta = 0{\cdot}05$; this value is also supported by tests on circular bridge columns by Chai *et al.* (1994). For the members of the structures studied, calculated β values according to the Park *et al.* (1987) empirical formula (included in the original IDARC code) were consistently lower than 0·01 in the beams and 0·08 in the columns, and were only used for the strength degradation parameter of the hysteretic model (not for calculating the damage indices). As the relative importance of the energy term depends heavily on the value of β, a second value of 0·25 was also used in further studies; this value generally corresponds to poorly detailed RC members, although no specific calibration has been reported so far for this case.

4.5.3. *Discussion of results*

Shown in Fig. 4.3 are the damage indices calculated for the members of the frame structure, under the most critical and the least critical input motion (see Table 4.1). The calculated maximum/minimum ratios for the seven motions varied from as low as 1·2 to as high as 3·0, with a reasonably similar distribution of damage indices along the height of the building.

With regard to the contribution of the energy term (see equation 4.3) to the value of the damage index, it was found that for the typical value of $\beta = 0{\cdot}05$ it was very low for all seven records considered. As shown in Table 4.1, energy accounts for up to only 9% in the beams and 5% in the columns for all the records studied; it must, moreover, be pointed out that in the vast majority of members the energy term accounts for 2–4% of the *D*-value, and the maxima listed in the table refer to some cases where the value of the damage index is quite low (0·10 or less), hence not critical for the structure. Of course, if β is increased to 0·25, the contribution of the energy term becomes approximately 5 times the previous value, that is 20–40%, as is also shown in the table.

It is therefore clear that, at least for adequately designed and detailed structures, and for loading histories that are typical of earthquake motions

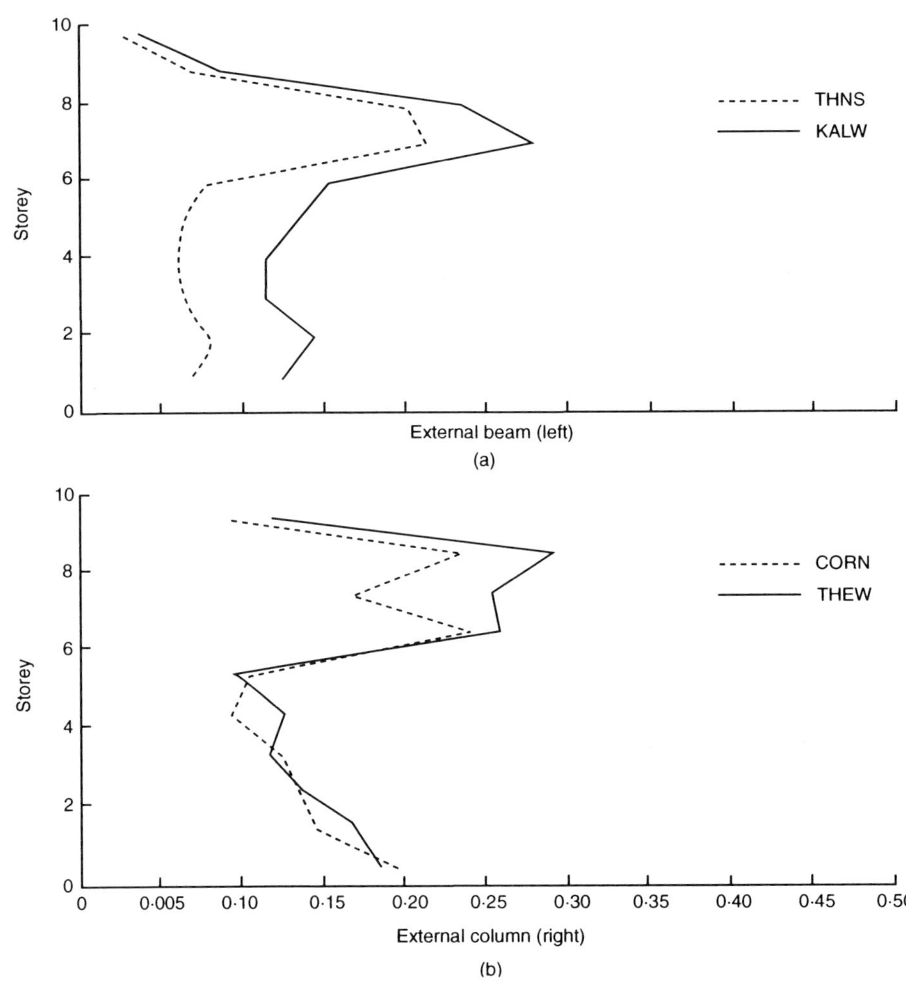

Fig. 4.3. Maximum and minimum values of Park–Ang damage indices calculated for the members of the frame structure: (a) beams; (b) columns

Table 4.1. Contribution (%) of energy term in calculated damage indices for frame structure

Input motion	Duration: s	Scaling factor	Beams $\beta = 0{\cdot}05$	Beams $\beta = 0{\cdot}25$	Columns $\beta = 0{\cdot}05$	Columns $\beta = 0{\cdot}25$
El Centro S00E	26·5	0·93	7·7	36·5	4·7	22·2
Thessaloniki N30E	12·0	2·17	7·9	37·7	4·9	23·0
Thessaloniki N60W	12·0	2·66	7·6	36·1	4·5	21·2
Corinth N35E	19·4	1·09	7·8	37·1	4·3	20·4
Corinth N55E	17·8	1·32	8·4	39·7	5·2	24·8
Kalamata N10W	8·5	1·24	8·6	40·9	4·7	22·3
Kalamata N80E	7·5	1·17	8·3	39·2	4·8	23·0

Table 4.2. Calculated damage indices in the beams of the frame structure for single and double El Centro record

Storey No.	El Centro + El Centro			El Centro			Calculated ratios		
	Ext. left	Ext. right	Int.	Ext. left	Ext. right	Int.			
(i)	(ii)	(iii)	(iv)	(v)	(vi)	(vii)	(ii)/(v)	(iii)/(vi)	(iv)/(vii)
1	0·376	0·359	0·552	0·147	0·143	0·218	2·56	2·51	2·53
2	0·281	0·315	0·417	0·121	0·131	0·166	2·32	2·40	2·51
3	0·213	0·155	0·287	0·109	0·102	0·152	1·95	1·52	1·89
4	0·131	0·111	0·137	0·097	0·089	0·133	1·35	1·25	1·03
5	0·161	0·151	0·225	0·129	0·125	0·174	1·25	1·21	1·29
6	0·164	0·185	0·249	0·148	0·124	0·197	1·11	1·49	1·26
7	0·478	0·471	0·695	0·199	0·204	0·297	2·40	2·31	2·34
8	0·393	0·399	0·593	0·188	0·158	0·245	2·09	2·53	2·42
9	0·112	0·134	0·159	0·081	0·067	0·132	1·38	2·00	1·20
10	0·052	0·019	0·047	0·032	0·019	0·031	0·63	1·00	1·52

recorded in earthquake-prone areas such as southern Europe and California, damage of RC members is controlled by peak deformation, rather than by low-cycle fatigue (as expressed by the energy term in equation 4.1). As a consequence of this, the duration of the record does not play the paramount role usually presumed in many previous studies with regard to damage potential; in fact it was found that the highest damage indices were calculated for the Thessaloniki N30E (duration of 12 s) or the Kalamata N10W (duration of 8·5 s) records, rather than for El Centro (duration of 26·5 s). This, of course, is due to the fact that peak response is usually not affected by duration, while energy is. It is also pointed out that the hysteresis model used in the present study takes some account of the effect of damage on the post-yield behaviour, since strength at a specified displacement level depends on the energy dissipated up to that level; despite this feature, the effect of duration was not significant for the records studied.

It is interesting to see what happens in the case of records with very long effective duration, such as the artificial one involving twice the strong part of El Centro considered in this study (which was found to be more damaging than actually recorded motions with long duration, such as that of Mexico City, 1985).

The calculated damage indices for this record are given in Table 4.2, together with the corresponding values calculated for the actual El Centro record (effective duration of 26·5 s). The contribution of the energy term for the artificial record (when the typical value of $\beta = 0{\cdot}05$ was used) ranged from 1·4–5·4% in the columns and from 3·3–7·6% in the beams, which means that the maximum values are essentially the same as in the actual ('single') El Centro record (see Table 4.1). This should be attributed to the fact that peak response has increased by approximately the same amount (up

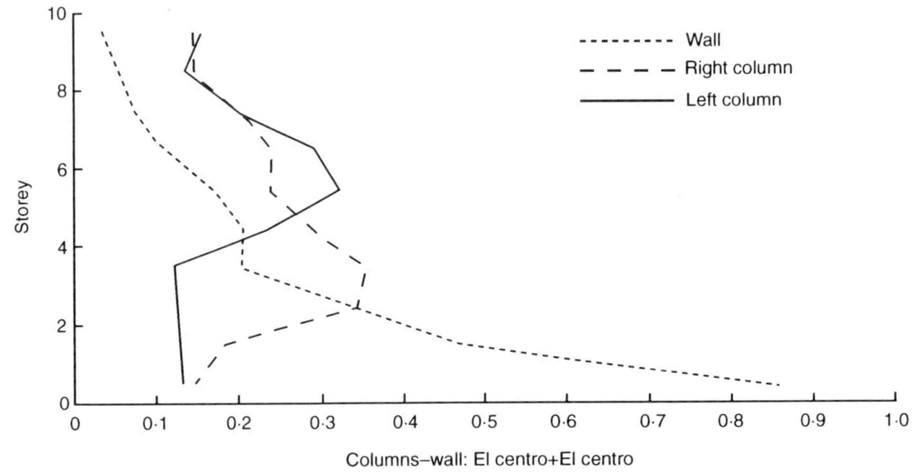

Fig. 4.4. Calculated Park–Ang damage indices for the vertical members of the dual structure subjected to two consecutive El Centro records

to about 150% as shown in the last columns of Table 4.2) as the energy dissipation, hence the relative contribution of each term in equation 4.1 has remained essentially the same.

A further point worth mentioning is that applying a second 26·5 s pulse of El Centro to the already damaged structure causes a rather non-uniform distribution of damage, as clearly seen by the ratios shown in Table 4.2, which vary from as high as 2·6 to values close to unity, and also from Fig. 4.4; this is a consequence of the trend of inelasticity to concentrate at the weakest parts of a structure, which has long been reported in the literature.

4.6. Conclusions

(*a*) A large amount of work has been done to date in the field of seismic damage measures for RC structures, and all possible approaches appear to have been explored, albeit not to the same depth. All indices suggested to date have been formulated so as to yield $D = 1$ at failure, but much less work has been done with regard to defining intermediate damage stages, in particular the threshold of damage and the threshold of irreparable damage (compare Fig. 4.1). This is clearly a direction which future research should concentrate on, if damage indices are to be used for practical purposes, that is for the design of new structures or the retrofit of existing ones.

(*b*) Since failure under seismic loading invariably involves some amount of cycling in the inelastic range, proper calibration of the damage indices requires a large amount of cyclic test data, carried out under similar, well-defined conditions (analogous to standard testing of concrete or steel strength and ultimate deformation under monotonic loading). As the available experimental database is still relatively limited, and the parameters influencing damage and failure are numerous, it is no surprise that the scatter associated with most currently available indices is very significant. The foregoing strongly point to the need to account properly for the probabilistic nature of damage indices, which for practical design purposes should be reduced to the determination of appropriate partial safety factors.

(*c*) Most of the available damage indices are based on a macroscopic approach, whereby flexural, shear and bond mechanisms are treated together, and corresponding analytical models are either purely flexural (based on curvatures) or account for the other two damage mechanisms in an approximate and/or empirical way. These should explain to a certain extent the large scatter mentioned previously. It might be worthwhile directing future research towards developing combined indices with distinct terms describing damage due to each mechanism; this has been attempted for combined flexure and shear, but with very limited calibration (for monotonic loading only) and limited analytical implementation in practical models.

(*d*) For well-detailed RC members, combined indices such as the widely used one by Park–Ang, appear to be dominated by the (monotonic) ductility term, while energy plays only a marginal role, even in (extreme) cases of earthquakes with very long duration of strong motion. This has significant implications with regard to design or assessment of structures, since the much simpler and far better calibrated ductility factor might be used as the main parameter for estimating structural damage.

(*e*) Despite their weaknesses, damage indices are a powerful tool that might be incorporated in future design and redesign procedures, with the potential to lead to more efficient and economical solutions. In fact, design methodologies based on the use of damage indices have already been suggested by some investigators (Park *et al.* 1987, Chung *et al.* 1988, Stone and Taylor, 1994). In addition to the previously discussed problems, a fundamental difficulty arising in such procedures is the appropriate definition of a global damage index for the entire structure, which in some cases can not be expressed simply as a weighted average of member indices, as has been suggested to date.

5. Selected case studies

5.1. Introduction

The rules applied for the proportioning and detailing of RC structures have a direct bearing upon their seismic response. Therefore, modern seismic design codes use detailed principles and application rules (collectively called provisions) to control the inelastic response of RC structures to ground motions at the design level and beyond. Extensive numerical and experimental studies are typically performed in order to calibrate the code provisions, so that the desired inelastic structural response is obtained. A few studies of this type are presented in this chapter.

The case studies in section 5.2 refer to two 10-storey plane structures, a 3-bay frame and a 2-bay dual (frame–wall) system, with columns proportioned in flexure according to eight different capacity design approaches. Other than for the column flexural design, the CEB (1985) model code for seismic design is applied for the proportioning and detailing of all members in bending and in shear. The alternative capacity design approaches examined either follow specific modern design codes (e.g. the Eurocode 8 for DCM, or the Greek code), or are more generic (e.g. following equation (2.12) of Chapter 2) or even quite sophisticated in concept and implementation. The eight so-designed pairs of structures, with their columns proportioned to match as closely as feasible the strength demands from the corresponding capacity design approach, are subjected to non-linear dynamic analyses of their response to several historic ground motions, normalized to the design ground motion intensity. For the structures designed according to two of the eight capacity design approaches, the non-linear analyses are repeated at twice the design motion intensity.

The case studies of section 5.3 focus on structures designed according to Eurocode 8. Twenty building structures in 3D, ranging from 3-storey to 12-storey, are considered, designed to the three different ductility classes (DC) of Eurocode 8 (corresponding to three alternative approaches to achieving the desired seismic performance). All these structures are designed according to both column capacity design approaches provided by Eurocode 8: the more strict conventional format of equation (2.12) and the relaxed one of equation (2.14). Twelve of the twenty structures, designed following the relaxed capacity design approach of equation (2.14), are then analysed non-linearly in the time-domain in one of the two horizontal directions, under four artificial, spectrum-compatible ground motions applied at the design motion intensity and at twice the design intensity (section 5.3.4). In section 5.3.5 twelve structures in both their capacity design variants, are also subjected to non-linear dynamic analyses, in both horizontal directions, under the same four artificial motions applied separately in each direction at 1·0, 1·5 or 2·0 times the design motion intensity. The level of non-linear modelling is about the same as that used in section 5.2, but much simpler than that of section 5.3.4. In this case, the effect of the participation of the slab to the tension flange of the beam is also examined, as a potentially important factor for control of the inelastic response.

In all the non-linear response analyses of section 5.3, columns are designed not only on the basis of the strength demands of the two alternative Eurocode 8 approaches, but also following the detailing rules and minimal measures of the corresponding ductility class. They possess, therefore, considerable overstrength beyond the corresponding uniaxial strength demands; even more so as they are proportioned for biaxial bending under simultaneous action of the two horizontal earthquake components, factored according to Eurocode 8. Accordingly, the results of their seismic response analyses reflect all sources of overstrength and portray the expected (in the average sense) seismic performance of structures designed according to Eurocode 8. In this respect these results are not directly comparable in a quantitative sense to those of section 5.2: the response to the design ground motion intensity predicted in section 5.2 can be considered to correspond roughly to something between those computed in section 5.3 for 1·0 and 1·5 times the design motion intensity. Moreover, the case studies of section 5.3 attempt to assess at the same time all measures provided by Eurocode 8 for the control of the inelastic response, including those for member detailing, and not just the provisions for the column capacity design in bending.

5.2. Influence of column capacity design method on the response of 10-storey plane frame and dual structures

5.2.1. Scope and objectives

There are at least nine distinct factors (Kappos, 1997a) that influence the capacity design of columns, and a rigorous procedure should account for all these factors, without being overly conservative. The aim of the case study presented here (Kappos, 1997a) was to assess the influence of the capacity design procedure on the performance of columns and of the structure as a whole, and to explore the possibility of developing new procedures which might combine lower cost of construction with increased reliability with regard to seismic loading; only 10-storey RC frames and dual systems with regular configuration are addressed. The study is directly related to codes of practice, such as the new Eurocode 8 (CEN, 1995), and its results could be used towards increasing their scope by explicitly including methods based on time-history analysis; the latter is currently recognized as a design method by the Eurocode 8 (EC8), but no specific guidance on its use is given, except for the selection of input motions.

5.2.2. Design procedures

Two regular RC buildings were designed, one with a structural system consisting solely of frames, and one dual (wall and frame) structure. Due to symmetry and regularity in plan the two buildings were idealized as the plane structures shown in Fig. 5.1, for the purpose of analysis. It is pointed out that dual structures are sometimes exempted from explicit capacity design for flexure, since the presence of appropriately designed walls precludes the formation of sidesway mechanisms which jeopardize the overall stability of the building under seismic loading. Nevertheless, avoiding column hinging in these structures as well, is a goal of good seismic design for various reasons (see Penelis and Kappos, 1997).

As the main purpose of the study was to evaluate the effect of column capacity design procedure, the beams of the structure (and the wall in the dual system) were designed according to the CEB (1985) Model Code, for an effective peak ground acceleration $A_d = 0{\cdot}25\,g$ and the intermediate ductility level (DL II). The materials considered were C20/25 concrete and S400 steel. The same code was also used for the shear design of all members. This code ensures an acceptable seismic reliability without being too conservative with regard to hoop reinforcement requirements in columns, as the case was found to be with EC8 (Kappos and Athanassiadou, 1997). The flexural design of columns was carried out using different capacity design procedures, as described in the following sections. Hence, the resulting structures differed only in the column longitudinal reinforcement, and on a few occasions in some column cross-sections as

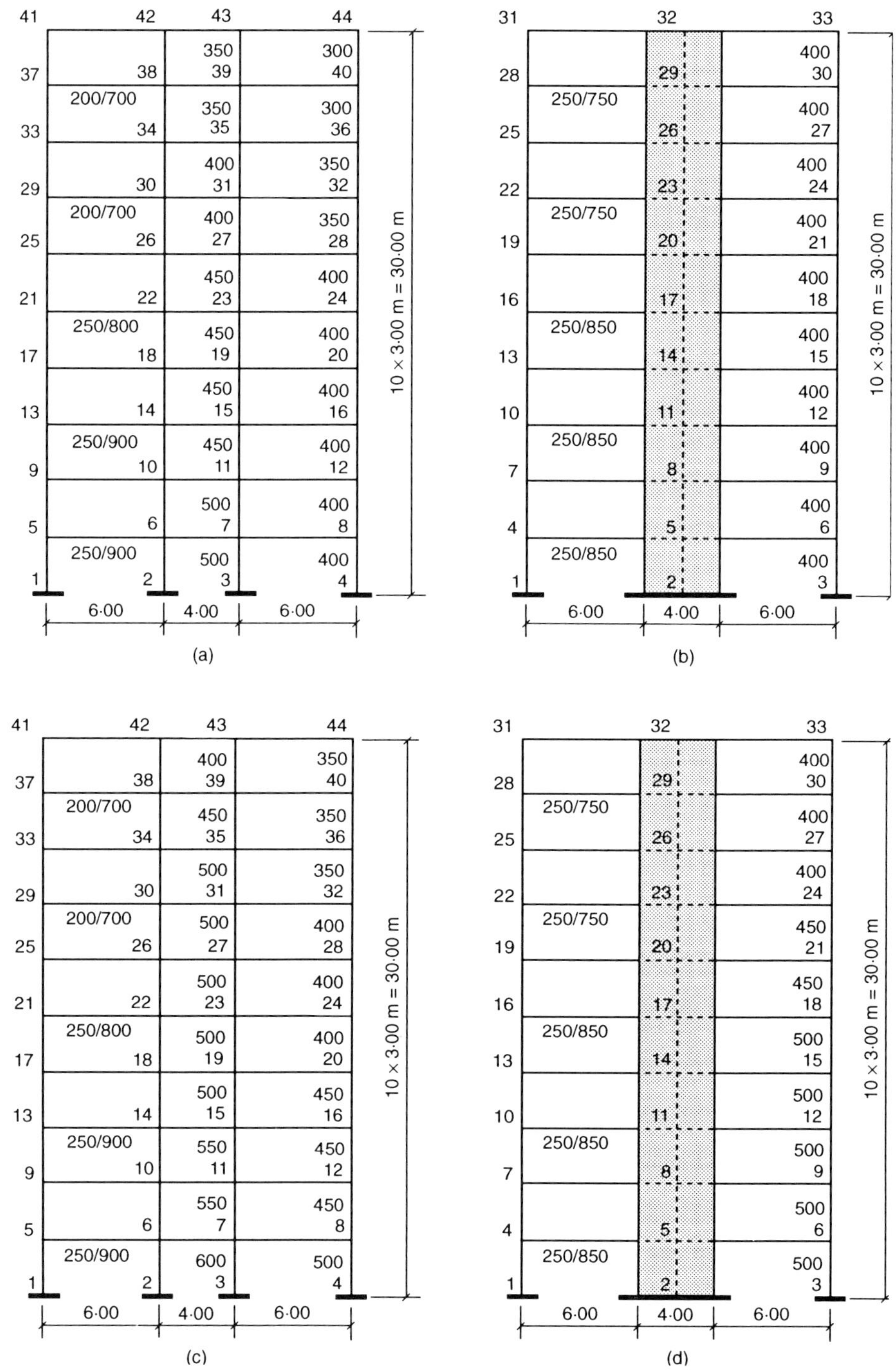

Fig. 5.1. Geometry of the structures analysed (wall thickness 250 mm): (a) frame structure designed for $\gamma_R = 1.0$; (b) dual structure designed for $\gamma_R = 1.0$; (c) frame structure designed according to new procedure; (d) dual structure designed according to new procedure

well (to avoid using excessive amounts of longitudinal reinforcement). In the case of EC8 design, an additional structure satisfying all requirements for columns was also designed and assessed.

5.2.2.1. Design methods used. Both buildings were designed using different existing methods for the protection of their columns, i.e. different procedures for increasing column moments (and shears) with respect to the values calculated from analysis for the seismic combination of actions. The methods considered included the following.

(*a*) The 'traditional' capacity design procedure adopted by various American and European codes, involving factoring beam strengths at a beam–column joint by appropriate overstrength (γ_R) factors and providing columns with a sum of resistances exceeding that of beams (equation (2.12)); the check is carried out using column action effects derived from the seismic combination (earthquake + reduced gravity load). The γ_R values used were 1·00 (with and without an additional dynamic magnification factor ω_D), and 1·50, a reasonable upper limit for practical design, as may be inferred from previous studies (Dolce and Evangelista, 1990; Colangelo, Giannini and Pinto, 1995).

(*b*) The Eurocode 8 (CEN, 1995) procedure of the a_{CD} factor (effect of $\gamma_R \Sigma M_{ub}$) and δ (moment–reversal) factor (equation (2.14)); the latter accounts for the reduced possibility of hinge formation at the bottom of gravity load dominated beams. Two cases were considered, one in which only column flexural design was modified with regard to the reference structure, and one in which the full set of EC8 provisions was considered.

(*c*) The approximate method of doubling the seismic moments (M_E) in columns, which is analogous to the procedure suggested by some national codes, such as the UBC, for deriving design shears in beams and columns of low or intermediate ductility.

(*d*) A modified EC8 procedure (which drops the δ-factor), adopted by the new Greek Seismic Code, which is essentially the same as the one used in the New Zealand Code (with somewhat different overstrength factors). The main difference between this procedure and those of (*a*) and (*b*), is that it explicitly requires distribution of ΣM_{Rb} (equation (2.12)) to the columns in proportion to the seismic moments only (not the total column moments).

In addition to the foregoing, a novel procedure was applied involving design of columns on the basis of inelastic dynamic analysis, wherein columns are assumed to remain elastic, while beams are allowed to yield. The part of the procedure regarding flexural design may be summarized as follows.

(i) Design of the beams of the structure according to standard code procedures, for seismic as well as for gravity loading; design moments coincide with those calculated from analysis, and if required, some moment redistribution may be carried out with a view to optimizing beam design. If walls are present, their critical regions (typically including the first one or two storeys) are also designed using moments calculated from analysis.

(ii) Detailing of the flexural reinforcement of beams (and wall critical regions), taking into account minimum code requirements and convenience of construction (e.g. use of a limited number of bar diameters).

(iii) Selection of an appropriate set of input accelerograms, using the techniques described in some modern seismic codes, such as EC8; either artificial, spectrum-compatible, or actually recorded motions appropriately scaled to match the design spectrum, may be used (see also section 5.2.3.3 on input motions).

(iv) Construction of a model of the structure wherein beams are modelled as yielding elements, with their strength based on the reinforcement actually present (including that in the adjacent slab), and with due consideration of factors such as stiffness degradation. In the same model, columns, as well as portions of walls (when present) intended to remain elastic, are modelled as elastic members. With regard to initial stiffness assumptions, it is recommended that 40–50% of the gross section stiffness EI_g be used for beams and 75–80% EI_g for columns (Paulay, 1989; Kappos, 1986), to account for (pre-yield) cracking, which is different in each type of member.

(v) Time-history analysis of the model described in the previous step for the selected set of input motions; calculation of critical moment (M) and axial load (N) combinations for each column (and wall) critical section.

(vi) Design and detailing of columns (and walls); for a column subjected to biaxial loading, consideration of three combinations will be sufficient for most practical purposes, while for uniaxial loading two combinations will suffice. The foregoing are valid on condition that the axial load limitations imposed by EC8 are respected; these limitations should be checked using the minimum N (maximum compression) calculated in the time-history analysis.

5.2.2.2. Final designs. In designing the two buildings according to the various methods, care was taken to retain some common features in all designs (in addition to using the same beams). A key aspect in this regard is that the reinforcement ratio in columns was kept as uniform as possible, so that the ductility of the various designs (which is affected by this ratio, both directly with respect to the rotational capacity, and indirectly with respect to the shear stress that can be developed) remained essentially unchanged. As shown in Fig. 5.1, this resulted in increases in column dimensions of up to 100 mm; normally such differences can be easily accommodated in the structure without creating architectural problems. From the practical point of view, it has to be pointed out that the selected beam sections in the frame structure tend to be unrealistic (smaller depths should normally have been selected, with corresponding higher reinforcement ratios, and inevitably lower ductilities).

In applying the new methodology it was found that in the frame structure the critical design combination for the columns was typically the one involving maximum $|M|$ and the corresponding N (which was always very close to the maximum compression value), while in the dual structure the critical combination was typically the one involving a tensile axial load N (or minimum $|N|$ for compression) and the corresponding M (which was less than the maximum M). It is worth pointing out that if the M and N values corresponding to a certain envelope (maximum or minimum) value have to be stored during the time-history analysis, the respective storage requirements double in the case of uniaxial bending and triple in the general case of biaxial bending.

The study focused on the aspect of column design, hence the reinforcement in the wall of the dual structure was kept the same in all cases (standard capacity design, as specified in the New Zealand and CEB Codes, as well as in EC8).

5.2.3. Method of assessment

The seismic performance of the structures designed to the previously described existing and new capacity design procedures was then assessed by analysing the response of appropriate inelastic models (all members allowed to yield) to a series of input accelerograms, and comparing the resulting demands with the corresponding capacities.

5.2.3.1. Modelling aspects. It is now recognized (CEB TG III/6, 1994) that analysis of realistic RC structures for earthquake excitations inducing inelastic response can best (and possibly only) be carried out using member-type models, where, typically, an actual structural member coincides with one finite element of the model; this approach was adopted in this case study. Standard point hinge modelling was used for RC members, with phenomenological hysteresis laws governing the behaviour of each hinge (see also section 4.4.1). The analysis of the structures was carried out using DRAIN-2D/90 (Kappos, 1996), an extended version of the well-known DRAIN-2D program (Kanaan and Powell, 1973), including several new features and elements. Based on experience from previous studies (Kappos, 1986), the two-component model, which has only the ability to handle bilinear hysteresis but accounts for yield moment M_y — axial force N interaction, was used for the exterior columns where a significant variation of the axial load takes place during the seismic excitation, while the one-component model, which is able to handle more refined Takeda-type hysteresis laws, but does not account for M_y–N interaction, was used for the other members of the structures (beams, interior columns and walls). Accounting for M_y–N interaction in the columns was an important feature of the study, since it is one of the major parameters of the problem. Strength and stiffness data for the RC members were determined using a fibre model for the analysis of critical sections, wherein appropriate constitutive laws for steel, and for unconfined and confined concrete (Kappos, 1991) were used. According to the standard approach adopted in deterministic structural assessment, mean values of the materials ($f_{cm} \cong f_{ck} + 8$ MPa for concrete, and $f_{ym} \cong 1{\cdot}1 f_{yk}$ for steel) were used for calculating strength data. Negative yield moments in beams were estimated assuming some additional contribution from the slab reinforcement which increased the area of top bars by 12–16%; note that the slab considered is a lightly reinforced one (3 m span).

As the main weakness of point hinge models is the constant stiffness value assigned to the intermediate line element (whose behaviour remains elastic throughout the response history), analyses were carried out assuming a moderate amount of cracking in the RC members (see section 5.2.2.1), due to gravity loading, as well as to previous moderate seismic excitations; the amount of stiffness reduction in the beams was assumed to be twice that in columns, to reflect different cracking patterns under predominantly gravity loading (Kappos, 1986). If significant inelastic behaviour was detected during the first analysis (average member ductility factors exceeding values around 2·0), further analyses were carried out assuming fully cracked line elements ($EI_{ef} \cong M_y/\phi_y$, where M_y equals the yield moment and ϕ_y the corresponding curvature at the critical section). This is a rough, yet practicable way for dealing with the inability of lumped plasticity models to simulate the gradual decrease of stiffness in RC members. It is pointed out that if inelasticity does not spread more or less uniformly along the height of the structure (that is if weak storeys are present), the reduced stiffnesses in the second analysis ($EI_{ef} \cong M_y/\phi_y$) should only be used in the regions with significant yielding, otherwise the seismic effects on the structure might be underestimated (due to the very low overall flexibility).

5.2.3.2. Failure criteria. The possibility of failure in each member, as well as in each storey of the buildings, was checked by applying appropriate local, as well as global failure criteria.

Local failure can be attributed to either of the following mechanisms

(*a*) exceedance of the available plastic rotation capacity θ_p of an RC member, taking the adverse effect of high shear stresses into account

Table 5.1 Basic data for the input motions used

No.	Earthquake	Magnitude: M	Site	Component	Peak acceleration: g	Peak velocity: mm/sec	Normalization parameter	Effective duration: sec
1	Volvi, Greece 20 June 1978	6·5	Thessaloniki (city centre)	N30E	0·142	127	2·17	12·0
2	Volvi, Greece 20 June 1978	6·5	Thessaloniki (city centre)	N60W	0·154	167	2·66	12·0
3	Alkyonides, Greece 24 Feb. 1981	6·7	Korinth (city centre)	N35E	0·239	225	1·09	19·4
4	Alkyonides, Greece 24 Feb. 1981	6·7	Korinth (city centre)	N55W	0·286	246	1·32	17·8
5	Kalamata, Greece 13 Sept. 1986	6·2	Kalamata (city centre)	N80E	0·240	323	1·17	7·5
6	Kalamata, Greece 13 Sept. 1986	6·2	Kalamata (city centre)	N10W	0·273	237	1·24	8·5
7	Imperial Valley, California 18 May 1940	7·1	El Centro site	S00E	0·348	335	0·93	26·5

(*b*) development of a shear force exceeding the corresponding capacity of the member at the maximum ductility level; shear capacity was checked with respect to all possible failure modes, that is diagonal tension, diagonal compression (web crushing) and sliding shear.

A detailed description of the foregoing criteria may be found in section 4.4.3, and elsewhere (Kappos, 1991, 1997; Penelis and Kappos, 1997).

Global failure was assumed to coincide with storey failure; a dual criterion based on a limiting inter-storey drift of 2% and the simultaneous development of a sidesway collapse mechanism involving all vertical members has been adopted for assessing storey failure. Regardless of mechanism formation, a structure was assumed to have collapsed if the inter-storey drift at any location exceeds a limiting value of 3%.

It is noted herein that the foregoing criteria, although admittedly rough, are compatible with the level of sophistication of the analytical model, and make feasible an assessment of the seismic performance of structures, taking into account most of the principal factors that influence this performance.

5.2.3.3. Input motions. Although the scope of the case study does not include new procedures for selecting the design transient motion (this is a more or less open problem to date, in particular with regard to seismic loading), an effort was made to select an appropriate suite of base motions which are consistent with the design spectrum. To this purpose a procedure for scaling all motions to the spectrum intensity of the derived velocity design spectrum within an appropriate band of natural periods, as suggested by Kappos (1991), was applied.

Most of the available records from earthquakes that caused serious damage, including collapses and casualties, in Greece during the last twenty years were used, including the earthquakes of Volvi (1978), Alkyonides (1981), and Kalamata (1986). All these records (see Table 5.1) are characterized by the fact that they come from surface earthquakes with small epicentral distances (these are the typical destructive earthquakes in Greece); hence the selected set of records can be considered as generally representative for the design of structures in Greece. For comparison purposes, the well-known S00E component of the El Centro 1940 record was also included in the study; these seven motions were also used in the design of the structures to the new procedure outlined in section 5.2.3.2. A further set of four Greek records (Argostoli 17.1.83, N–S and E–W components; Argostoli 23.3.83, E–W component; Edessa 1990 N–S component) and the well-known S16E component of the Pacoima Dam (1971) record, representing an event significantly different from those commonly occurring in the area under consideration, were used in the case of the structures designed to the new methodology, in order to study their reliability when the earthquake input is different from that used in the design.

All records were normalized to fractions of the intensity of the design earthquake ($A_d = 0{\cdot}25$ g), using the modified Housner technique suggested by Kappos (1991), whereby the scaling factors are derived by considering the areas under the velocity response spectra of the actual records and of the design earthquake (code spectrum), between appropriately defined bounds of the natural period, corresponding to the fundamental modes of the structures under consideration. In order to assess the behaviour of all structures under the same intensity of seismic excitation, a single scaling factor was calculated for each record, corresponding to the period range from 0·6–1·9 s. The lower bound is the estimated fundamental period of the dual system assuming moderately cracked sections, while the upper bound is the period of the frame structure assuming fully cracked sections and rather significant yielding. The resulting normalization factors (see Table 4.1) are higher than the ratios of the design acceleration (0·25 g) to the peak acceleration of each record.

5.2.4. *Discussion of results*

A selection of the most representative results from the extensive parametric study involving the 16 different (as far as columns are concerned) models and the 7 + 5 records scaled to various intensities, are presented in the following sections.

5.2.4.1. Response to 0·25 g earthquake. Shown in Figs 5.2(a) to 5.17(a) are the distributions of plastic hinges in the various structures subjected to the most critical among the motions considered (the one producing the smallest safety margin with respect to failure); open circles correspond to yielding at one face of a member, and solid circles indicate

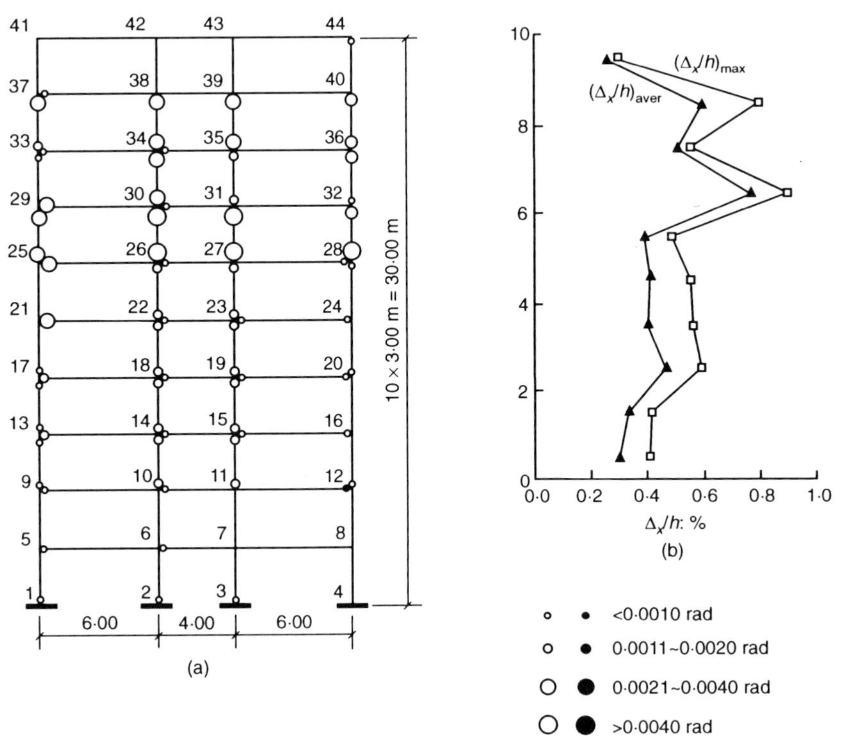

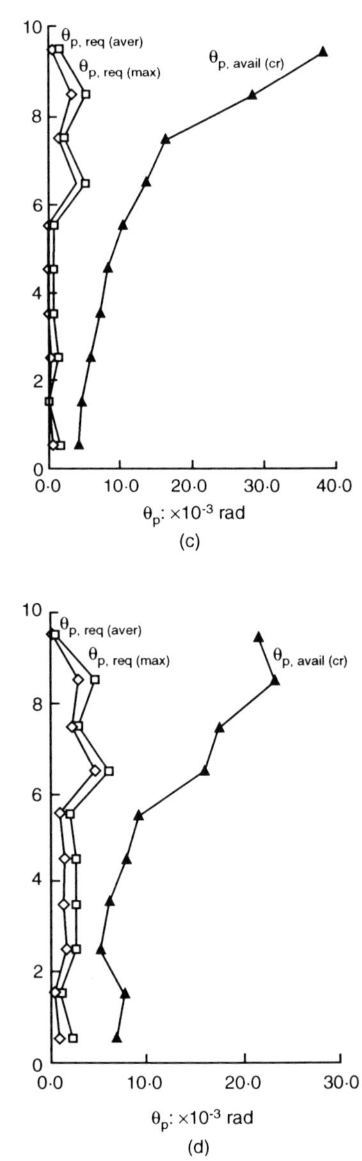

Fig. 5.2. Frame structure designed for $\gamma_R = 1.0$: (a) plastic hinge distribution for the most critical motion; black dots indicate yielding at both faces of a member, white dots yielding at one face only; (b) inter-storey drift ratios; (c) required and available plastic rotations in exterior columns; (d) required and available plastic rotations in interior columns

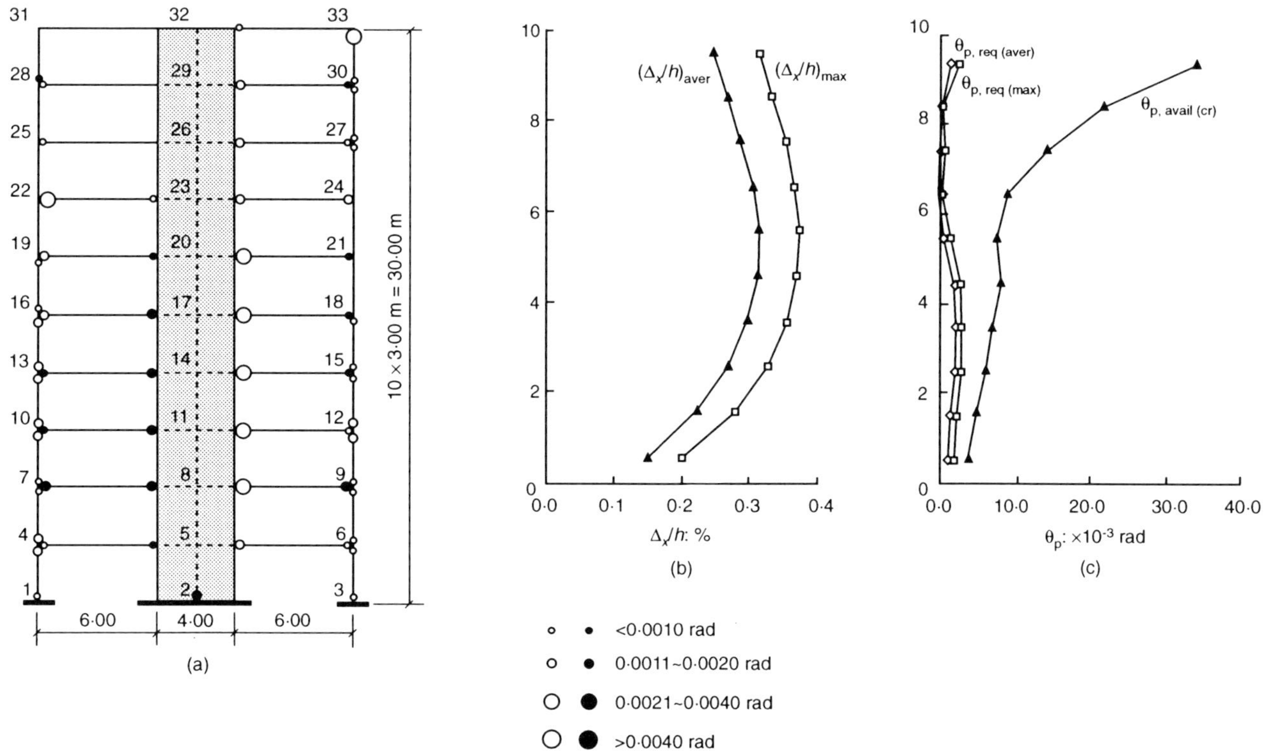

Fig. 5.3. Dual structure designed for $\gamma_R = 1.0$: (a) plastic hinge distribution for the most critical motion; (b) inter-storey drift ratios; (c) required and available plastic rotations in columns

yielding at both faces. Plotted in Fig. 5.2(c), (d), 5.3(c) up to 5.16(c), (d), 5.17(c) are the average (over seven motions in all cases, except in Figs 5.16 and 5.17, where twelve motions are considered) and the peak values of plastic rotations $\theta_{p,req}$ recorded during the dynamic analysis, together with the corresponding available rotational capacities $\theta_{p,avail}$ calculated on the basis of the previously mentioned local failure criteria for the most critical motion (conservative approach). Finally, Figs 5.2(b) to 5.17(b) show peak and average values of the inter-storey drift ratios; the black squares in 5.16(b) and 5.17(b) correspond to the average drifts for the five motions of the second set.

As expected, designing the columns for the minimum capacity design requirements ($\gamma_R = 1$, no ω_D factor) did not prevent hinge formation in these members under the design earthquake (Fig. 5.2(a), 5.3(a)). Indeed, at the seventh and the ninth storey of the frame the potential for a column sidesway mechanism was detected (Fig. 5.2(a)). However, due to the fact that yielding does not occur in all hinges at the same time, this mechanism is not as critical as might be inferred from the figure; this is manifested by the relatively low drift values, which do not exceed 0·9% (Fig. 5.2(b)), being well below the limit of 2%. Hinges form at all columns of the lower storeys of the dual structure (Fig. 5.3(a)), but in this case the wall remains in the elastic range (except at its base) and no sidesway mechanism can form. Hence, if the performance criterion is to avoid column hinging (above the ground level), this capacity design procedure is obviously inadequate. On the other hand, the plastic rotation requirements in the columns are quite moderate in both structures, the maximum values not exceeding 0·006 rad. It is clear from Figs 5.2(c), (d) and 5.3(c) that the capacities are well above the corresponding demands, the minimum 'safety factor' $\theta_{p,avail}/\theta_{p,req}$ being at least 1·9 (third-storey interior column of frame structure). Checks of the column shear capacities have shown that the corresponding safety factor is

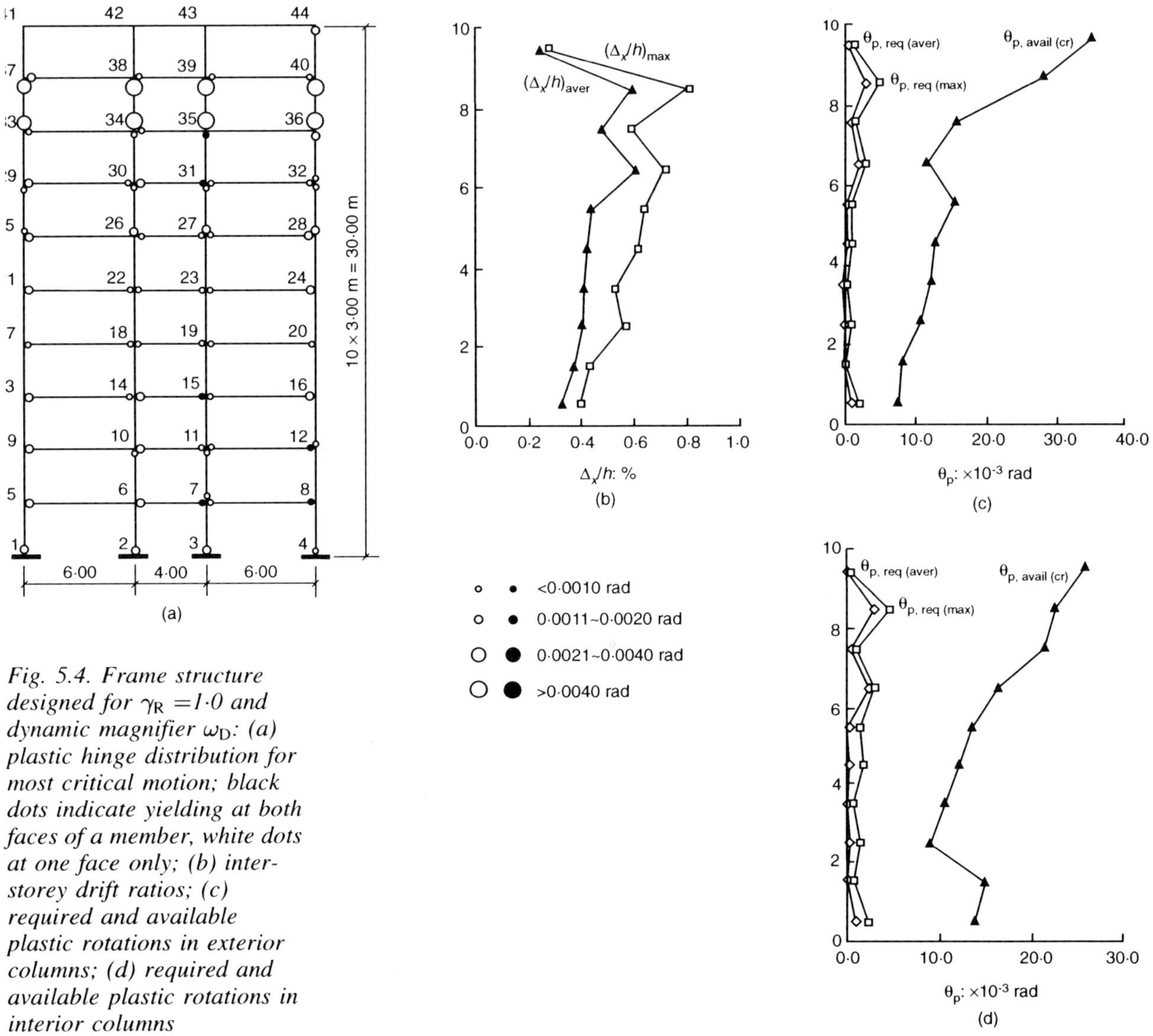

Fig. 5.4. Frame structure designed for $\gamma_R = 1{\cdot}0$ and dynamic magnifier ω_D: (a) plastic hinge distribution for most critical motion; black dots indicate yielding at both faces of a member, white dots at one face only; (b) inter-storey drift ratios; (c) required and available plastic rotations in exterior columns; (d) required and available plastic rotations in interior columns

2·5 or more for all columns. It should be remembered that the hoop reinforcement, on the basis of which local failure criteria were evaluated, was significantly lower than that required by EC8. Hence, if the performance criterion is to ensure an adequate ductility (and shear capacity) of the columns under the design earthquake, it can be claimed that even this minimal capacity design might be adequate in this respect.

Increasing the overstrength factor either directly ($\gamma_R = 1{\cdot}50$), or indirectly by introducing the ω_D factor of the New Zealand and the CEB codes, did not significantly alter the previously described picture, as column hinging did occur, sporadic in the former case (Figs 5.6(a) and 5.7(a)), but in the latter one (Figs 5.4(a) and 5.5(a)) even a potential column mechanism appeared in the same storeys as in the structure designed for $\gamma_R = 1{\cdot}0$. The difference was that the required plastic rotations were now lower and the corresponding minimum safety factors were 2·3 for both designs, but not for the same structure; the columns of the frame were the critical ones in the former case and those of the dual system in the latter, as shown in Figs 5.4(c), (d), 5.5(c), 5.6(c), (d) and 5.7(c). Moreover, the safety factors were generally higher than for $\gamma_R = 1{\cdot}0$. Therefore, increases of γ_R up to about 50% do not preclude hinge formation in columns, as also demonstrated in previous studies (Dolce and Evangelista, 1990, 1991).

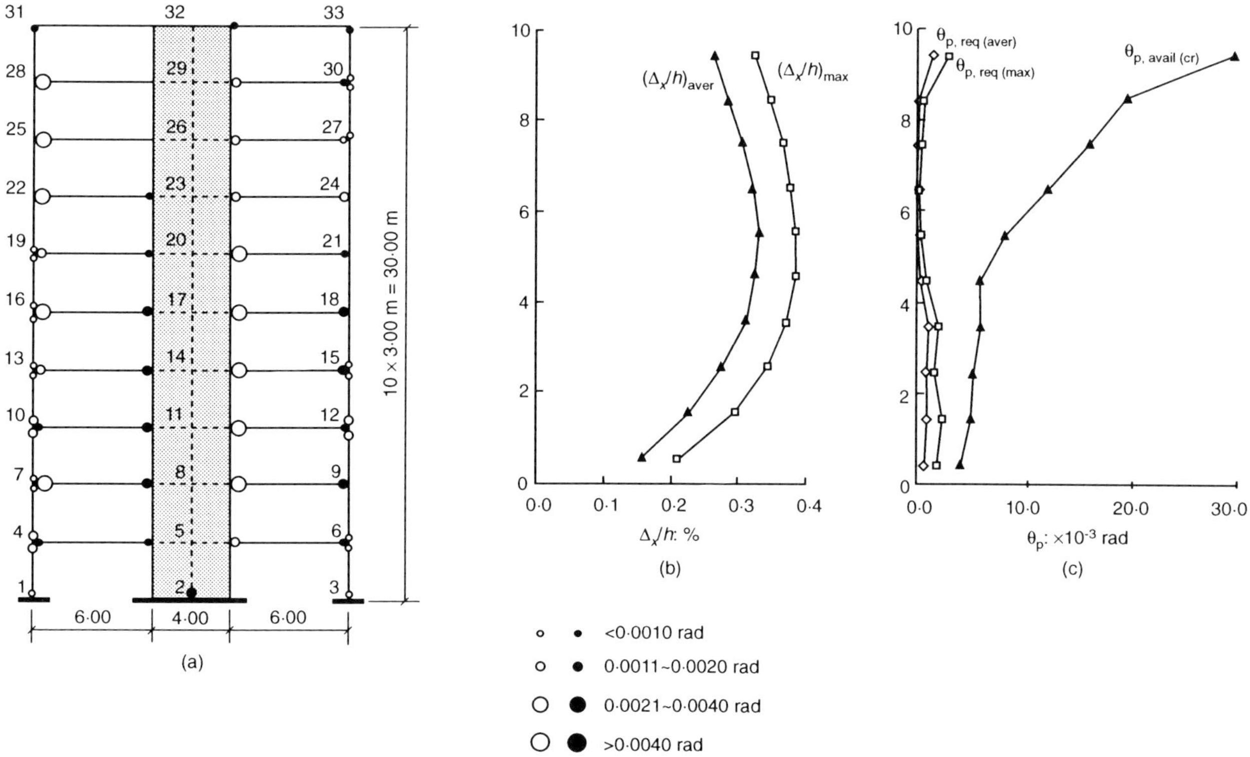

Fig. 5.5. Dual structure designed for γ_R = 1·0 and dynamic magnifier ω_D: (a) plastic hinge distribution for the most critical motion; (b) inter-storey drift ratios; (c) required and available plastic rotations in columns

The overstrength factor adopted by EC8 for DCM structures is γ_R = 1·2 and in the upper storeys it is somewhat reduced through the δ-factor. Hence, it was no surprise to discover that the behaviour of columns in the structures designed only to the EC8 provisions for column strength was very similar to that of the previously discussed designs, with a potential mechanism appearing at the seventh storey of the frame (where a column taper occurs), as can be seen in Fig. 5.8(a), and a minimum safety factor with regard to rotational capacity of 2·8. Introducing the remaining EC8 provisions for columns resulted in some improvement, with a minimum safety factor of 5·0 (in the frame); however, a potential mechanism then appeared at the ninth storey (Fig. 5.10(a)). A more detailed discussion of structures designed according to all EC8 provisions (for beams, columns and walls) and for all three ductility classes, can be found in Kappos and Athanassiadou (1997).

The modification to the EC8 capacity design of columns introduced in the new (1995) Greek Seismic Code consists of dropping the δ-factor and distributing the sum of beam moments ΣM_{ub} according to the seismic (not the total) column moments, a similar procedure to that used in the NZS3101 code. This procedure led to a further improvement in that no potential mechanism appeared in either structure (Figs 5.12(a) and 5.13(a)); the minimum safety factor recorded at the base of the ground-storey columns was only 1·9 (this became the most vulnerable part of the frame structure). Nevertheless, factors of 5·5 or more were recorded in all other storeys (see Fig. 5.12(c), (d)).

Among the conventional capacity design methods, the best protection of columns was achieved in cases where the seismic moments were doubled (regardless of beam strengths). For regular structures such as those studied

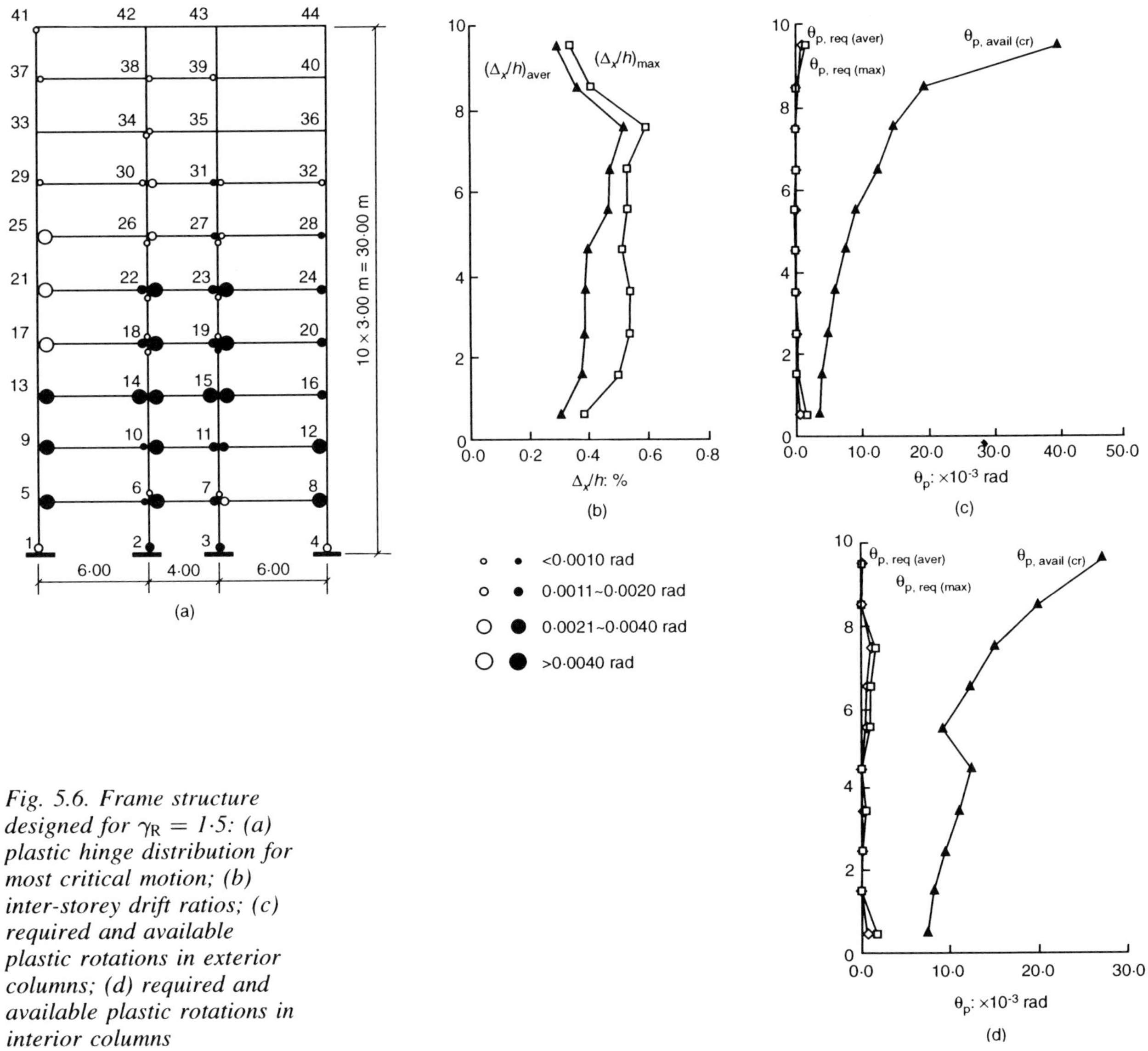

Fig. 5.6. Frame structure designed for $\gamma_R = 1{\cdot}5$: (a) plastic hinge distribution for most critical motion; (b) inter-storey drift ratios; (c) required and available plastic rotations in exterior columns; (d) required and available plastic rotations in interior columns

herein the foregoing is a conservative (and possibly not cost-effective) procedure. Column hinging was essentially restricted to the base of the exterior columns of the ground storey, in both the frame (Fig. 5.14(c)) and the dual system (Fig. 5.15(a), (c)). The pertinent safety factor was at least 11·3, recorded at the ground-storey columns; some minimal plastic rotations (0·001 rad, or less) were recorded at a few intermediate storeys of the frame (see Fig. 5.14(c), (d)) and at the top, where column hinging is permitted by the design. Hence, this procedure may be deemed to satisfy (although not literally) the performance criterion of no hinge formation in columns other than those of the first and the top storey.

Both structures designed to the proposed new methodology performed exactly as intended, even when they were subjected to the set of five input motions that had not been previously considered in the design process. As shown in Figs 5.16(a) and 5.17(a), column hinging appeared only at the base of the frame structure, and at the base and the top of the dual system; for one of the second set of five motions, one column section at the ninth storey just entered the inelastic range ($\theta_p = 0{\cdot}2 \times 10^{-3}$ rad). Given that inevitably only a few critical load (M, N) combinations were considered in design, the

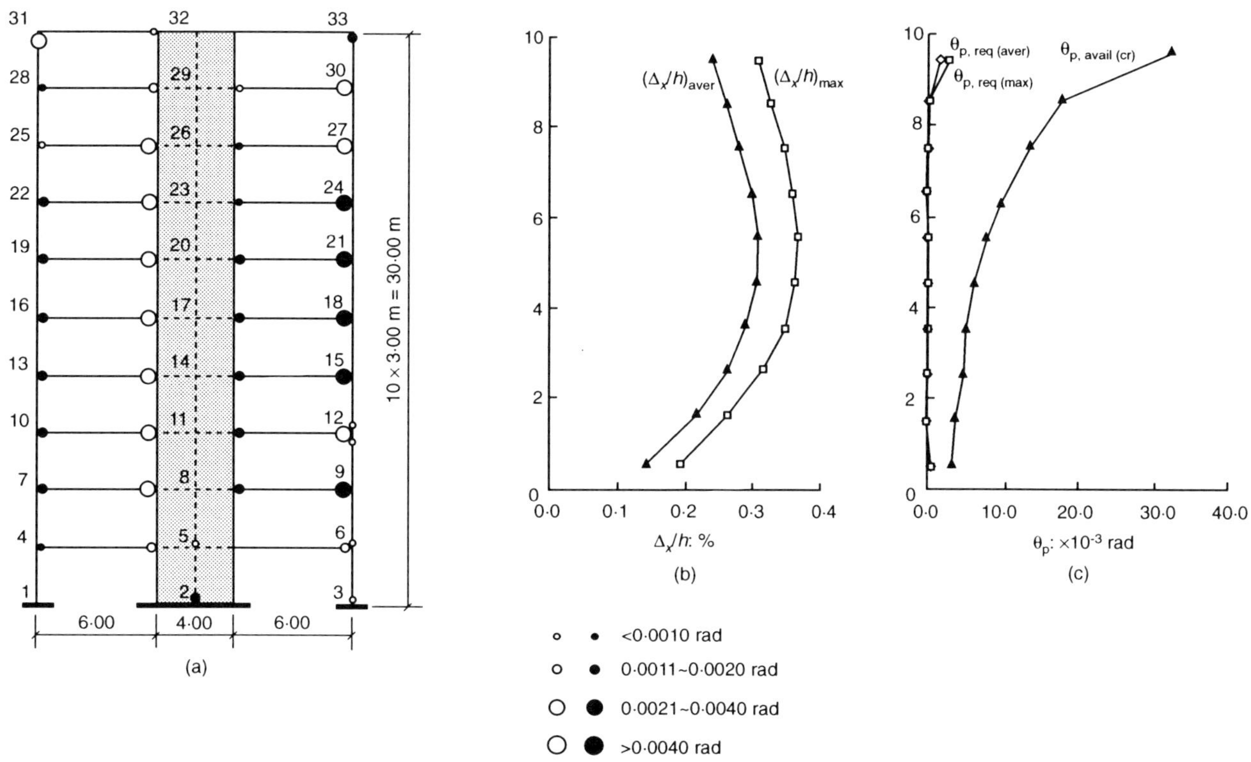

Fig. 5.7. Dual structure designed for $\gamma_R = 1{\cdot}5$: (a) plastic hinge distribution for most critical motion; (b) inter-storey drift ratios; (c) required and available plastic rotations in columns

foregoing appears to be a very adequate performance, clearly satisfying the requirements of all modern design codes. This improved performance was achieved by just a minor increase of material cost (column sections were up to 5 mm larger than those in the standard EC8 design), which can be more than offset if the confinement requirements in columns are significantly reduced in most storeys, on the basis that columns will remain elastic or very nearly so. A full calibration of this significant aspect of the proposed method clearly requires consideration of various design ductility classes, as well as of seismic actions higher than those corresponding to the design earthquake.

5.2.4.2. Response to 0·50 g earthquake. It was beyond the scope of this case study to discuss the definition of an earthquake action corresponding to the 'survival' limit state, wherein significant damage in a structure is considered as acceptable, so long as partial or total collapse and subsequent loss of life is avoided. Paulay and Priestley (1992) suggest that the probability of exceedance of the earthquake corresponding to the 'survival' limit state should be one-tenth that of the design earthquake, and the corresponding displacement ductility twice as large (e.g. 8, compared to a design ductility of 4). On the basis of these assumptions the 'survival' earthquake can be approximated by doubling the intensity of the design one. Hence, additional analyses for $A_{ef} = 0{\cdot}50\,g$ were run using the records of Table 4.1 with scaling coefficients equal to twice the values given in the eighth column. Only the limiting cases of capacity design for $\gamma_R = 1$ (no overstrength) and the new procedure will be discussed below; further discussions of the EC8 and CEB designed structures under both the serviceability and the survival earthquakes can be found in Kappos (1997b).

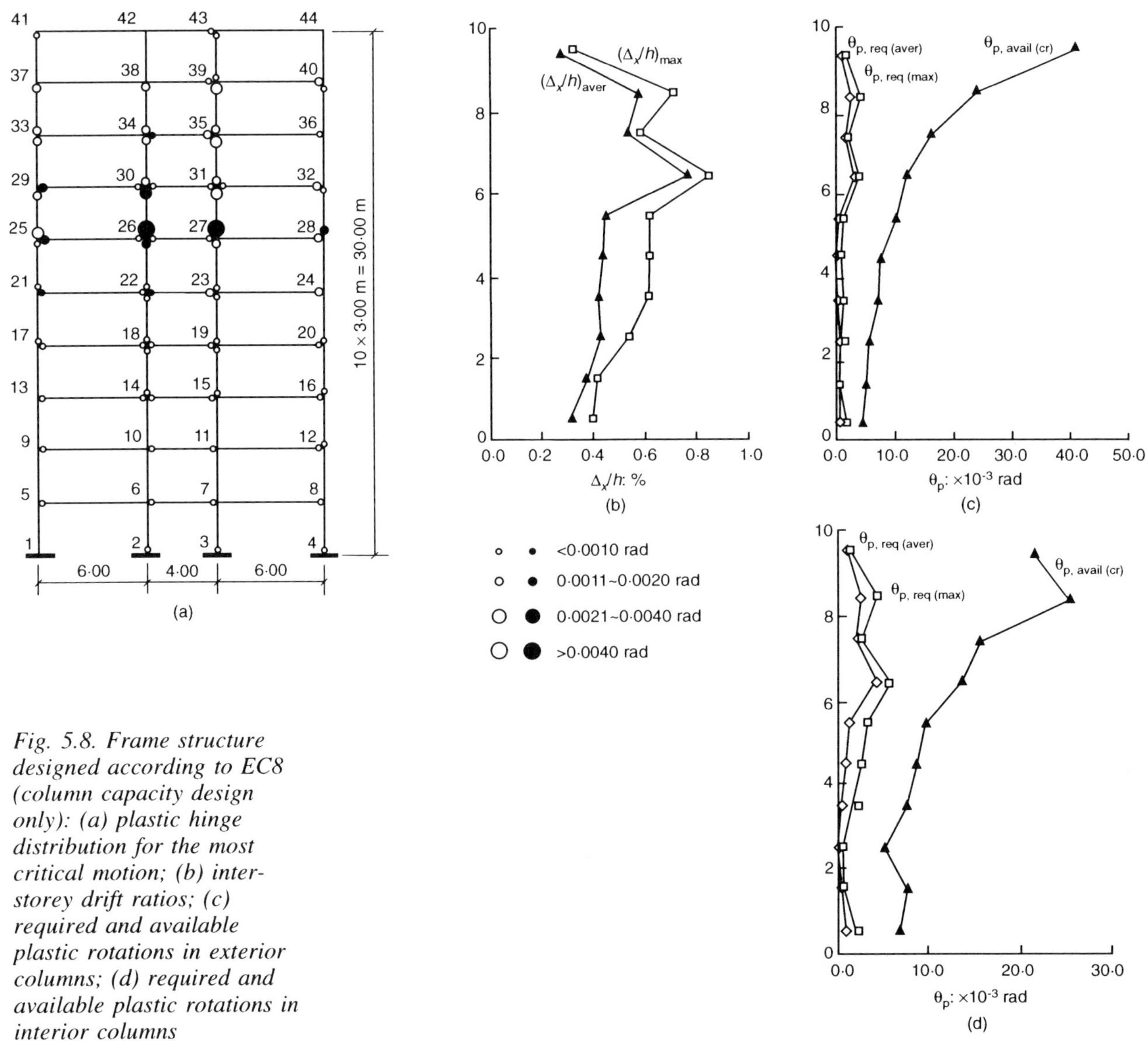

Fig. 5.8. Frame structure designed according to EC8 (column capacity design only): (a) plastic hinge distribution for the most critical motion; (b) inter-storey drift ratios; (c) required and available plastic rotations in exterior columns; (d) required and available plastic rotations in interior columns

Shown in Fig. 5.18 are the main response quantities calculated for the frame structure designed for $\gamma_R = 1$; based on the conclusions of previous studies using point hinge models (Kappos 1986, 1991), two different models of the structure were used, as discussed in section 5.2.4.1. As can be seen in Fig. 5.18(a), extensive column yielding was recorded in all but the top storey, resulting in very significant inter-storey drift ratios (Fig. 5.18(b)). On the basis of Fig. 5.18(a) and (b) it is clear that the global failure criterion suggests that the structure cannot withstand this earthquake. Moreover, failure is also suggested on the basis of the local criterion, as the required plastic rotations are higher than the corresponding capacities, in particular at the lower storeys, as shown in Fig. 5.18(c) and (d).

Mainly due to the fact that wall yielding was confined to the lower two storeys (Fig. 5.19(a)), the inter-storey drift ratios calculated for the dual structure designed for $\gamma_R = 1$ did not exceed 1·1%, even when the lower bounds on member stiffness were assumed (Fig. 5.19(b)). However, failure of columns at the lower two storeys should be expected, as indicated in Fig. 5.19(c). It is seen that capacity design ignoring beam overstrength cannot guarantee survival under the earthquake considered, even when dual structures with strong walls are involved.

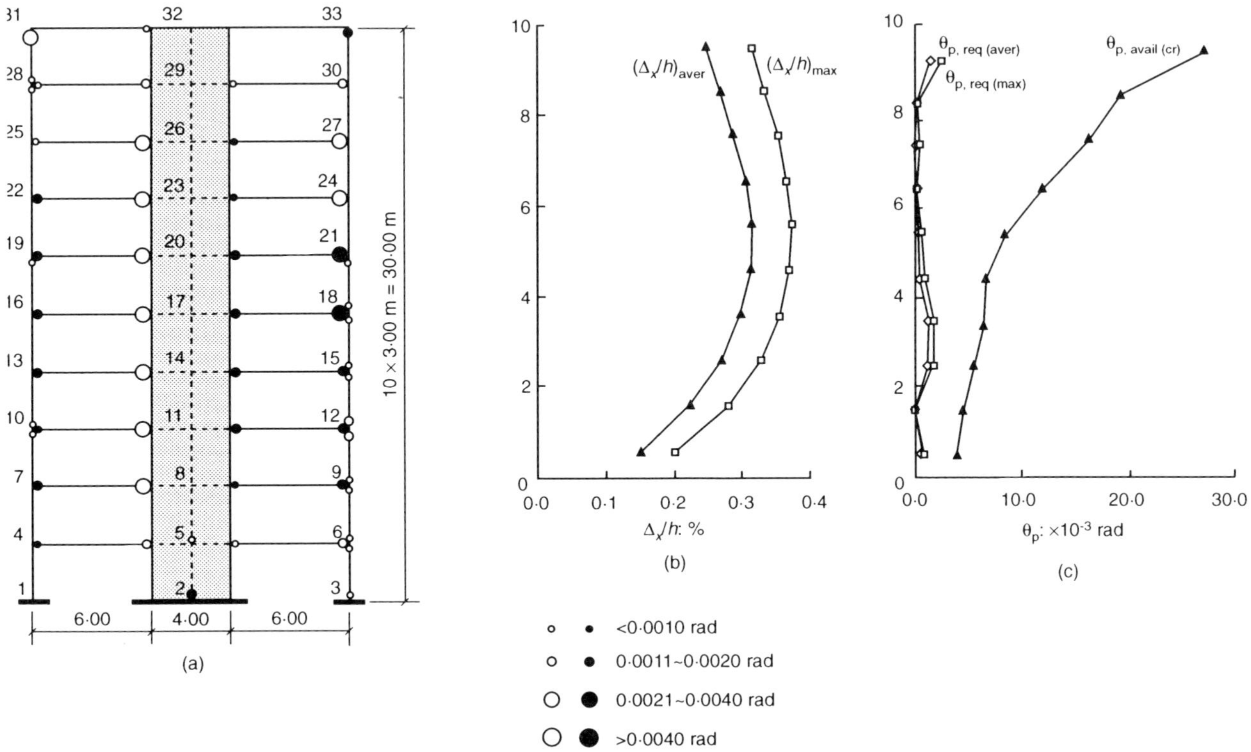

Fig. 5.9. Dual structure designed according to EC8 (column capacity design only): (a) plastic hinge distribution for the most critical motion; (b) inter-storey drift ratios; (c) required and available plastic rotations in columns

The foregoing behaviour should be compared with that of the buildings designed to the new methodology, shown in Figs 5.20 and 5.21. Even under an excitation (Pacoima S16E) that had not been considered at the design stage, yielding in columns for $A_{ef} = 0{\cdot}50\,g$ was minimal and very sporadic (Fig. 5.20(a)) with subsequent drifts up to about 1·6% and safety factors, with regard to exceeding the available rotational capacity (see Fig. 5.20(c), (d), black dots), of 2·6 or more. Yielding in the columns of the dual structure (Fig. 5.21) was also minimal, mainly concentrating at the top two storeys. The calculated drift ratios did not exceed about 1% (Fig. 5.21(b)) and the available rotational capacities were well above the corresponding requirements (Fig. 5.21(c)). The behaviour suggested by Figs 5.20 and 5.21 clearly points to the high seismic reliability of the structures designed using the new methodology.

5.2.5. *Concluding remarks*

The evaluation of various existing capacity design methods for columns presupposes the clarification of the related seismic performance criteria, which unfortunately is not achieved in most, if not all, current codes (EC8 not excluded). If the target performance is to avoid column hinging under the design earthquake at all storeys (except at the base and the top of the structure), then the present case study (in line with some previous ones) has clearly demonstrated that overstrength factors of 2·0 or more have to be introduced, even for the regular structures studied herein. If the target performance is to avoid formation of column sidesway mechanisms, the overstrength factors can be reduced (to some value higher than 1·5), while if the only requirement is to provide columns with sufficient ductility (in addition to adequate shear capacity) to withstand the design earthquake

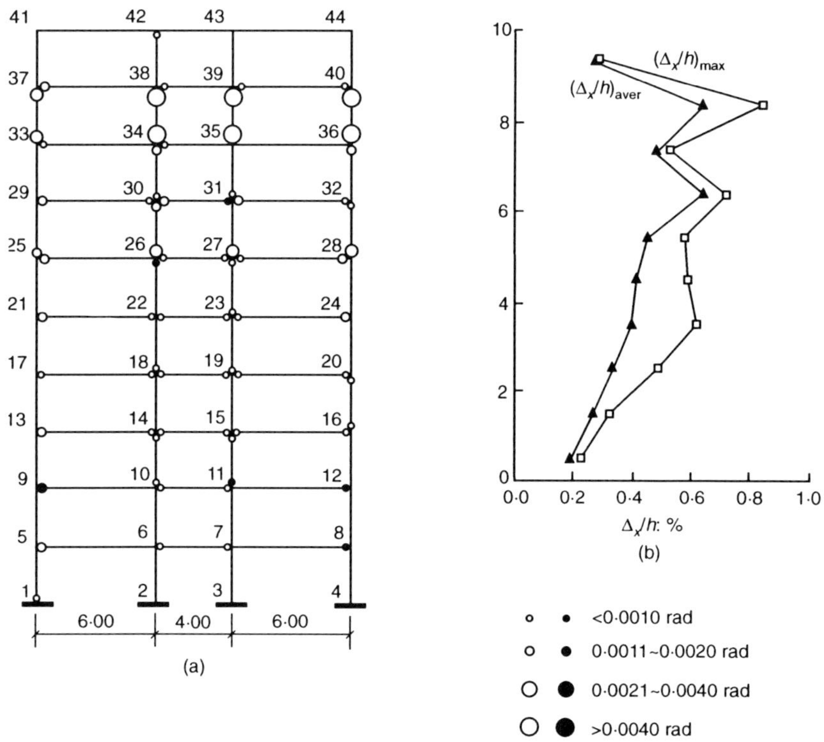

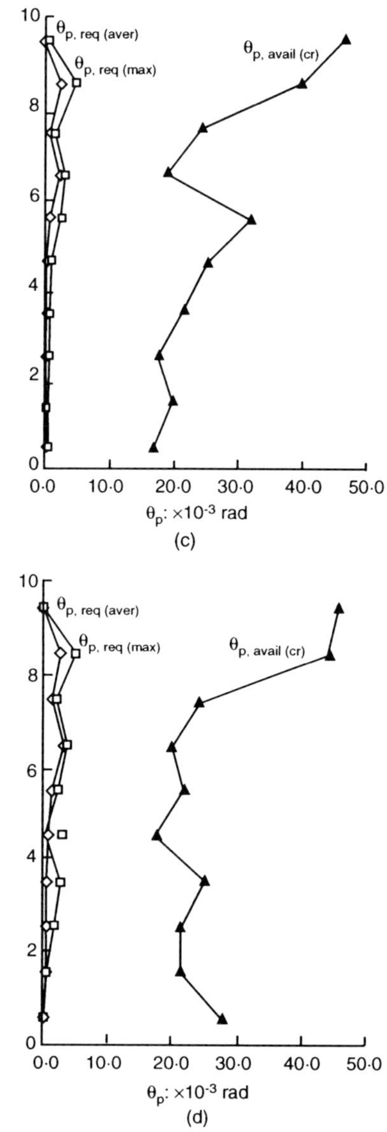

Fig. 5.10. Frame structure designed according to EC8 (all provisions): (a) plastic hinge distribution for the most critical motion; (b) inter-storey drift ratios; (c) required and available plastic rotations in the exterior columns; (d) required and available plastic rotations in the interior columns

without serious damage and/or collapse, then γ_R values slightly higher than 1·0 might be appropriate, at least for regular structures.

Among the various procedures studied, the very simple method of doubling the seismic actions in columns resulted in a particularly good performance (again the issue of regularity has to be borne in mind here), while the new procedure of capacity design led to a very satisfactory seismic performance, and may offer the possibility of reducing the cost of materials compared to that resulting from the code methods. The ultimate goal would of course be to strike an appropriate balance between flexural strength and ductility of columns (and all other members which do not form part of the primary energy dissipation system), with a view to optimizing the overall design.

5.3. Capacity design of multi-storey building structures

5.3.1. Scope and objectives

Modern seismic design codes, such as Eurocode 8 (EC8), impose limits on strength and deformation characteristics of RC structures to ensure the realization of pre-defined response and failure modes. To define new

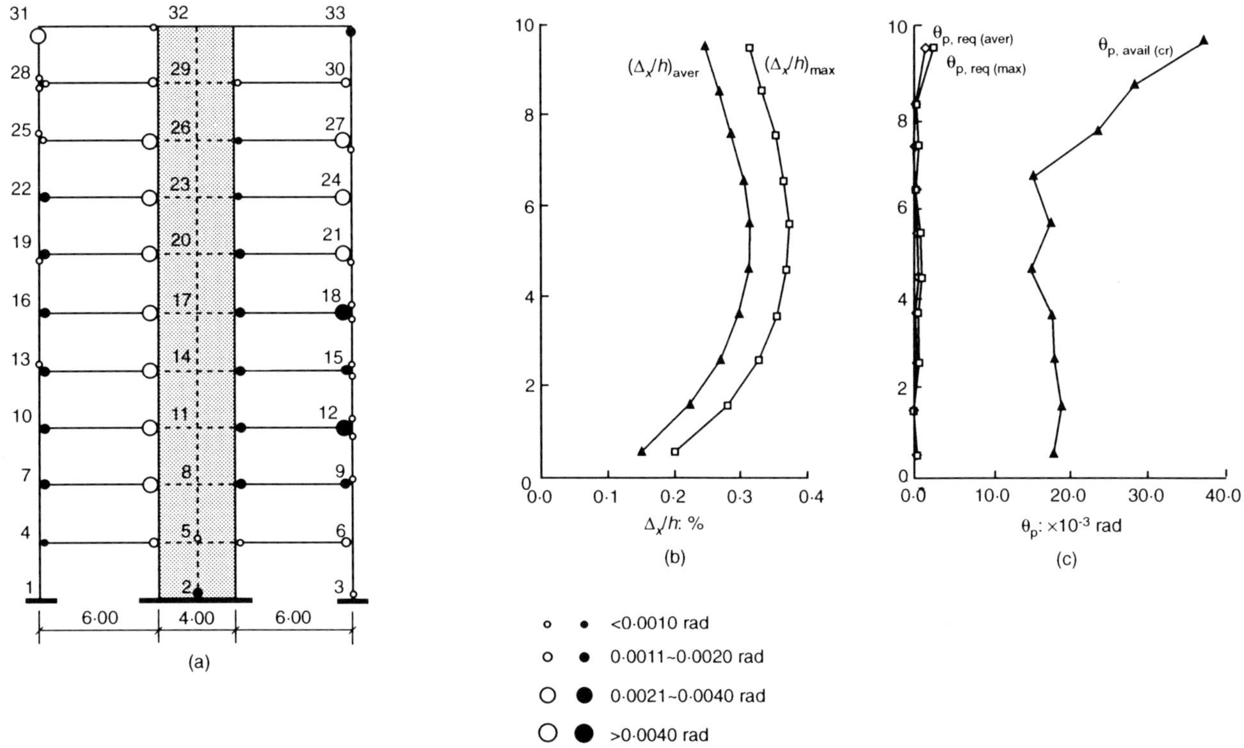

Fig. 5.11. Dual structure designed according to EC8 (all provisions): (a) plastic hinge distribution for the most critical motion; (b) inter-storey drift ratios; (c) required and available plastic rotations in columns

criteria and to refine existing ones, it is important that the interrelationship between the design and detailing conditions on the one hand, and the actual inelastic response characteristics on the other, is studied carefully. This objective may be served via two, possibly parallel, routes:

(*a*) analysis of a number of structures where one capacity design parameter is varied at a time
(*b*) analysis of a number of structures where different ductility class criteria have been applied.

The former option has the advantage of providing a unitary data point relating one capacity design parameter, e.g. shear overdesign or connection overstrength, to imposed demand and available supply of ductility. It has the disadvantage of studying fictitious structures that do not conform to any given set of design regulations, hence the structures are unrealistic. The latter option remedies this disadvantage by focusing attention on realistically designed and detailed structures. It has, however, the disadvantage that a number of criteria for design and detailing are varied concurrently, hence no unique parameter–response relationship is clearly established. It is believed, however, that both approaches are indeed complementary and should therefore be developed in tandem. It is through this dual approach that existing code regulations are assessed (via option (*b*)) and modified if need be (via option (*a*)).

Option (*a*) was followed in the case studies of section 5.2, in which alternative capacity design approaches are applied for the determination of the design bending moments M_{Sd} of the columns, and the columns are considered to be proportioned for a design resistance M_{Rd} exactly equal to

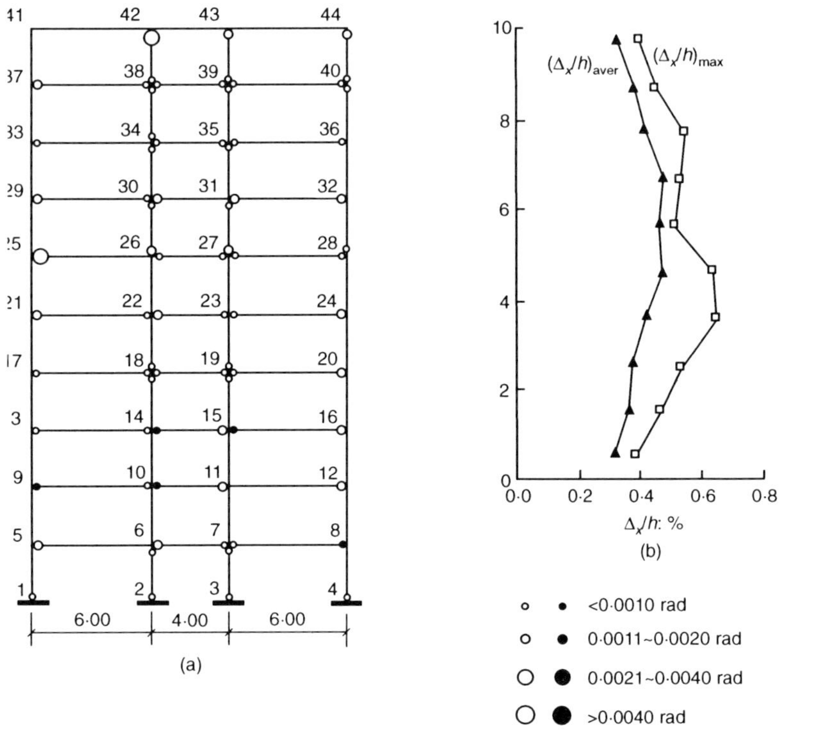

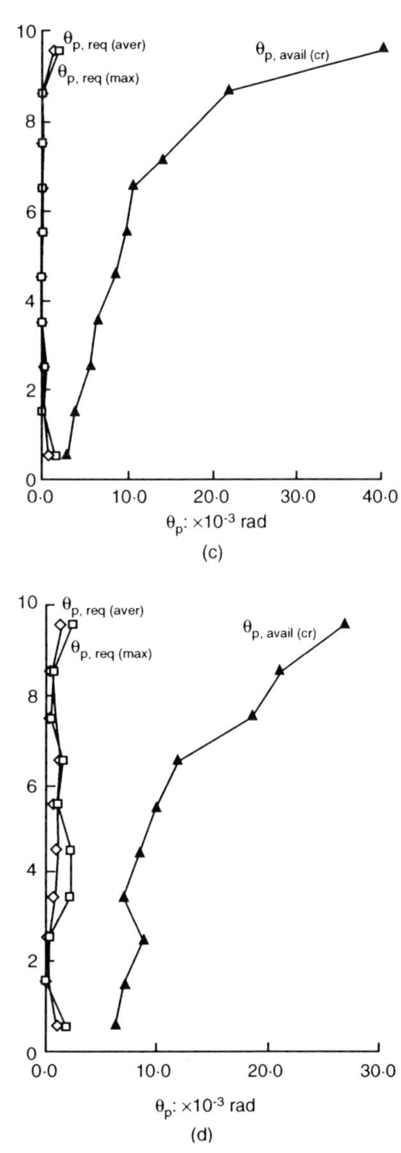

Fig. 5.12. Frame structure designed according to modified EC8: (a) plastic hinge distribution for the most critical motion; (b) inter-storey drift ratios; (c) required and available plastic rotations in exterior columns; (d) required and available plastic rotations in the interior columns

the corresponding M_{Sd}. In this way the effect of proportioning the columns for different capacity design determined action effects shows up more clearly, as it is not overshadowed by detailing, or other requirements which often exert control. In this section, option (*b*) is adopted and applied to a total of 26 multi-storey RC building structures which are fully designed and detailed according to Eurocodes 2 and 8.

In the following sections, the 26 building structures are described, their design according to alternative column capacity design approaches allowed by Eurocode 8 is presented, the input motions used for the inelastic seismic response analyses are described and, finally, the computed inelastic response of the structures to motions of up to twice the design level intensity is summarized and discussed.

5.3.2. Design of structures to different capacity design levels

5.3.2.1. Configuration and global design results of the structures. For the purposes of assessing the operationality of Eurocode 8, the effectiveness of its provisions for earthquake resistance, including those for capacity design, and the equivalence of the three ductility classes in terms of seismic

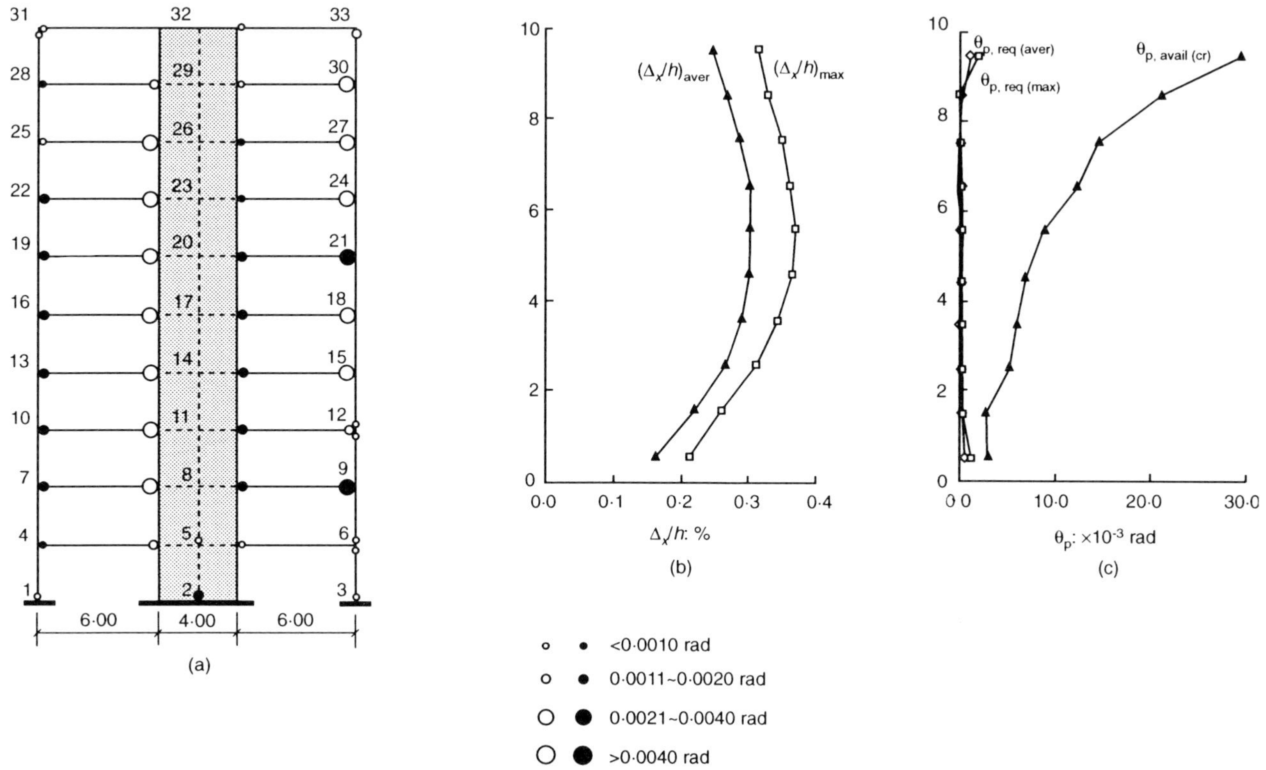

Fig. 5.13. Dual structure designed according to modified EC8: (a) plastic hinge distribution for the most critical motion; (b) inter-storey drift ratios; (c) required and available plastic rotations in columns

performance, a total of 26 RC buildings were fully designed according to Eurocodes 2 and 8 (Fardis, 1994). In these 26 buildings, columns were proportioned following the 'relaxed' capacity design approach of Eurocode 8, given by equation (2.14) of this Design Guide. Herein an alternative column design is also considered for the 26 buildings, following the more demanding general capacity design format of equation (2.12) in section 2.4.3.2, and the implications for the capacity of the columns are examined and discussed.

The 26 buildings of the pilot application of Eurocode 8 can be categorized into six groups depending on their structural configuration. In three of the groups the plan layout is similar in all storeys. These groups are called '4-storey', '8-storey' or '12-storey' structures, as the buildings have four, eight or twelve storeys, with constant storey height (see Figs 5.22 to 5.24). Another group has the same plan layout as the aforementioned three groups in seven of its eight storeys, while in the ground floor an irregularity in elevation is introduced by discontinuing the four 2nd-from-the-corner columns on the long (X) exterior sides and by increasing the first-storey height by 50% (see Fig. 5.25). This group is called 'irregular frame' structure. In another group, again with eight storeys, the four central columns are replaced by a core of two large channel-shaped shear walls, coupled in the short direction (Y) of the plan at all floors, and the solid two-way slabs on beams are replaced by a waffle-slab supported directly by the vertical elements. Moreover, the 2nd-from-the-corner columns on the long (X direction) sides are omitted in all storeys, as in the ground floor of the irregular frame structures. This group, shown in Fig. 5.26, is called 'core' structure. The last group is a 3-storey industrial building with high live load (5 kN/m^2 instead of the 2 kN/m^2 of the other groups) and long spans in

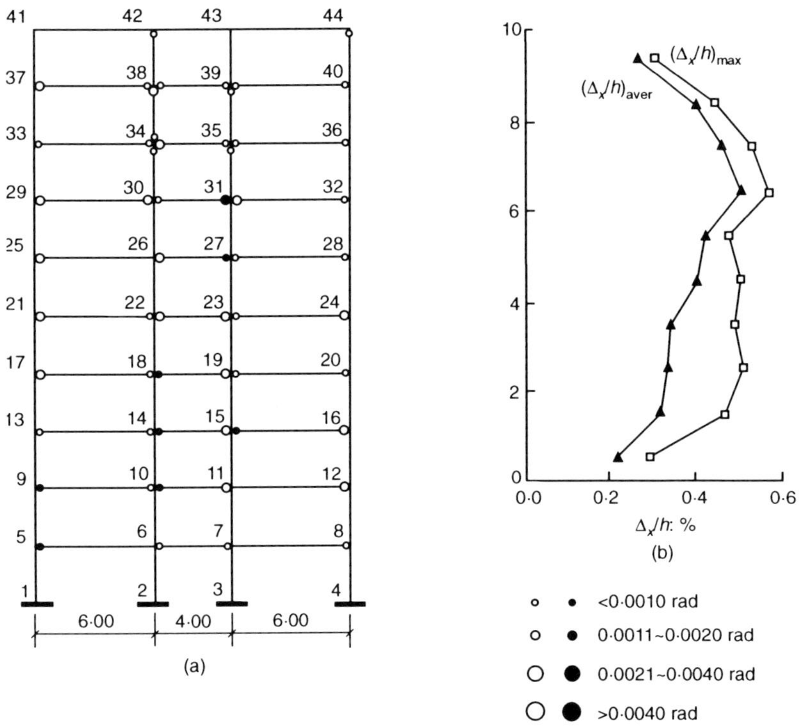

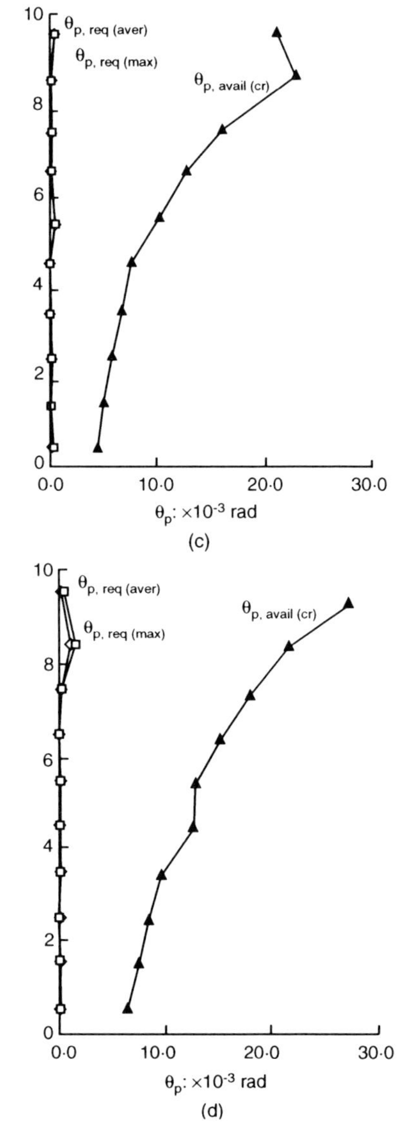

Fig. 5.14. Frame structure with columns designed for twice the seismic loading (2E): (a) plastic hinge distribution for the most critical motion; (b) inter-storey drift ratios; (c) required and available plastic rotations in exterior columns; (d) required and available plastic rotations in interior columns

direction Y (Fig. 5.27), so that gravity loads control the strength of the corresponding beams. This group is called '3-storey' structure. In all groups but this last one, a distributed permanent load of 2 kN/m^2 is considered to represent finishings and partitions at all floors. Such load is not considered in the group of 3-storey structures. For this group a line load equivalent to a weight of 0·4 kN/m^2 of vertical surface is considered for infill walls along all the beams of the perimeter, while the roof is considered to have only a permanent load of 3 kN/m^2 of horizontal area.

There are four buildings in each of the six groups, two of which are designed for a peak ground acceleration of 0·15 g and ductility classes L and M, and two for an acceleration of 0·3 g and for DCM and H. The groups of the 12-storey buildings and of the irregular 8-storey ones include a fifth structure, of DCM with design acceleration 0·3 g, for which the 'simplified modal response spectrum' (or 'equivalent static') method of linear-elastic analysis is used for the calculation of the design internal forces, instead of the 'multi-modal response spectrum' ('dynamic') method, used for the design of all other buildings. As shown in the third column of Table 5.2, the basic behaviour factor q_0 is taken as equal to 5 for all frame structures and

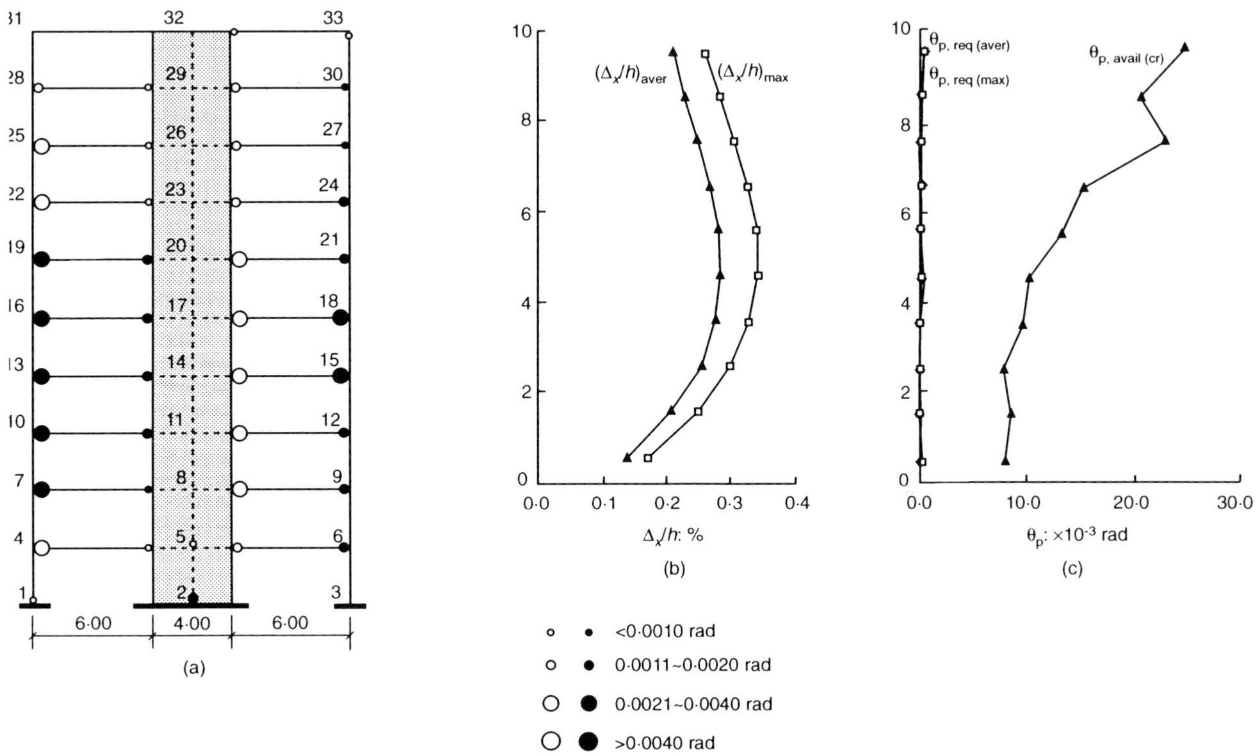

Fig. 5.15. Dual structure with columns designed for twice the seismic loading (2E): (a) plastic hinge distribution for the most critical motion; (b) inter-storey drift ratios; (c) required and available plastic rotations in columns

to 3·5 for the core structures. In the irregular frames the behaviour factor is reduced by 20%, over the value of q_0, as the k_R factor of Eurocode 8, Clause 1.3.2.3.2.1, for regularity in elevation is taken as equal to 0·8. In all other structures k_R is equal to 1·0. The k_w factor, reflecting the prevailing mode of behaviour is also taken as equal to 1·0 in the core structures, as the aspect ratio of the shear walls is such that their expected mode of behaviour and failure is in flexure, in both horizontal directions.

The concrete grade is C25/30 in all elements of the 26 buildings and the steel grade is S500 for both the longitudinal and the transverse reinforcement of all members.

Member cross-sectional dimensions of the 26 buildings are selected in general to be the same in each pair of buildings designed for the same ground acceleration but for different DC. Cross-sectional dimensions of columns and walls are kept constant in all storeys. With very few exceptions, beam cross-sectional dimensions are also kept the same in all storeys. To avoid significant overstrengths, associated with minimum steel ratio controlling member reinforcement, an effort is made to select the smallest possible member dimensions, within the limits set by the inter-storey drift control limitations, by the desire to keep second-order (P–δ) effects low and by the maximum allowable steel ratios of members (especially beams). In the 4-storey buildings, column cross-sectional dimensions are controlled by the anchorage requirements of beam bars in the joints and range between 0·4 m and 0·5 m. In the other types of building, column dimensions are dictated by the verification of the bottom section of the ground-storey columns in biaxial bending with axial force (according to Eurocode 8, at this section moments are magnified using the same capacity design magnification factors as at the top section of the ground storey).

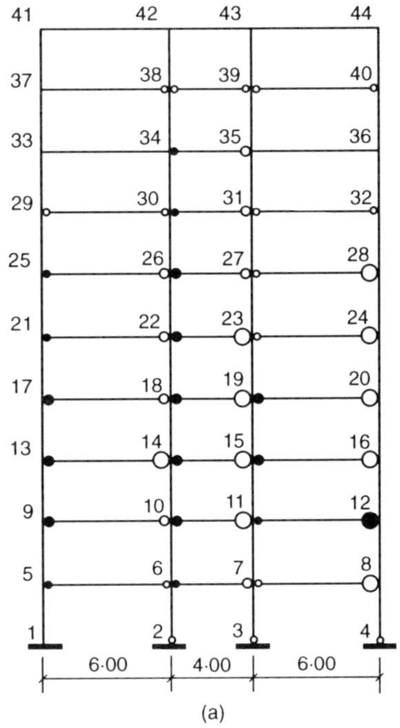

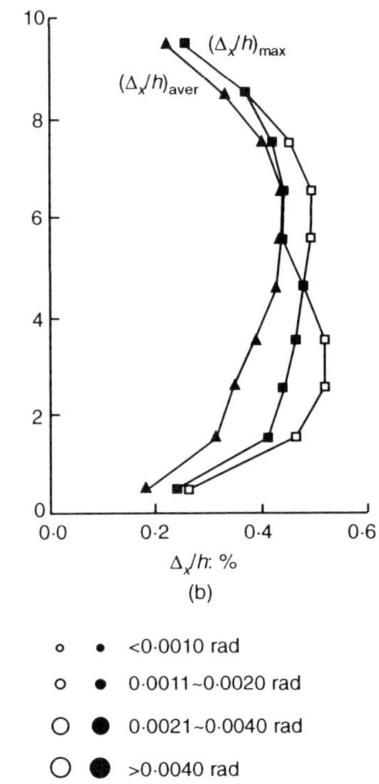

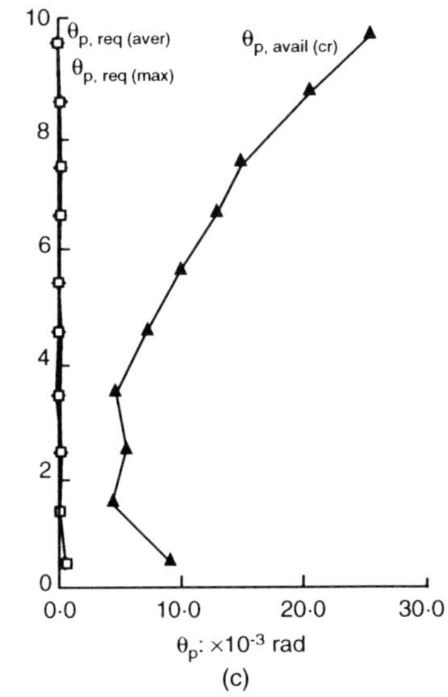

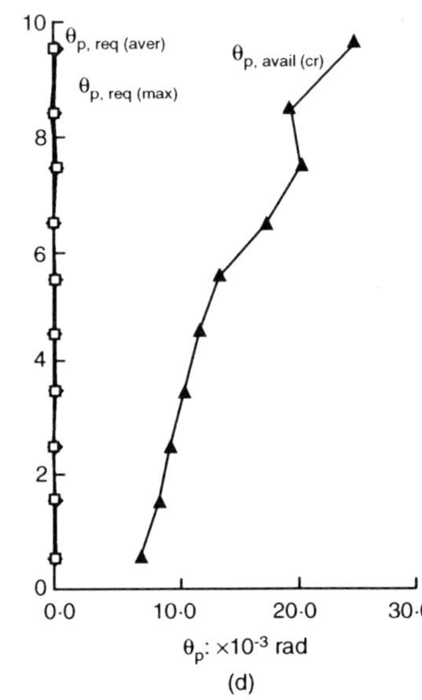

Fig. 5.16. Frame structure designed according to new procedure: (a) plastic hinge distribution for the most critical motion; (b) inter-storey drift ratios; (c) required and available plastic rotations in exterior columns; (d) required and available plastic rotations in interior columns

In the 'irregular frame' structures, the selected depths of the ground-storey beams supporting the discontinued columns are larger than the corresponding ones in the storeys above (0·8 m instead of 0·6 m), to concentrate the gravity load and the effects of the vertical component of the seismic action (taken into account in design in this particular case) on these larger ground-floor beams, and to avoid distributing it equally to the beams of all the storeys.

Table 5.2 summarizes global design results of the 26 structures: the fundamental period, the design base shear coefficients V_b/W and the top and the maximum inter-storey drift ratio δ/h. These global response results are computed on the basis of the uncracked stiffnesses of the members, as required by Eurocode 8. Table 5.2 also lists the total volume of concrete and weight of steel in the lateral load resisting part of the structure (i.e. excluding the slabs), and the distribution of steel between beams and columns and between longitudinal and transverse reinforcement (separately for beams and columns). These material quantities refer to proportioning of columns according to the relaxed capacity design approach of Eurocode 8, given by equation (2.14). For given design ground acceleration, the DC has

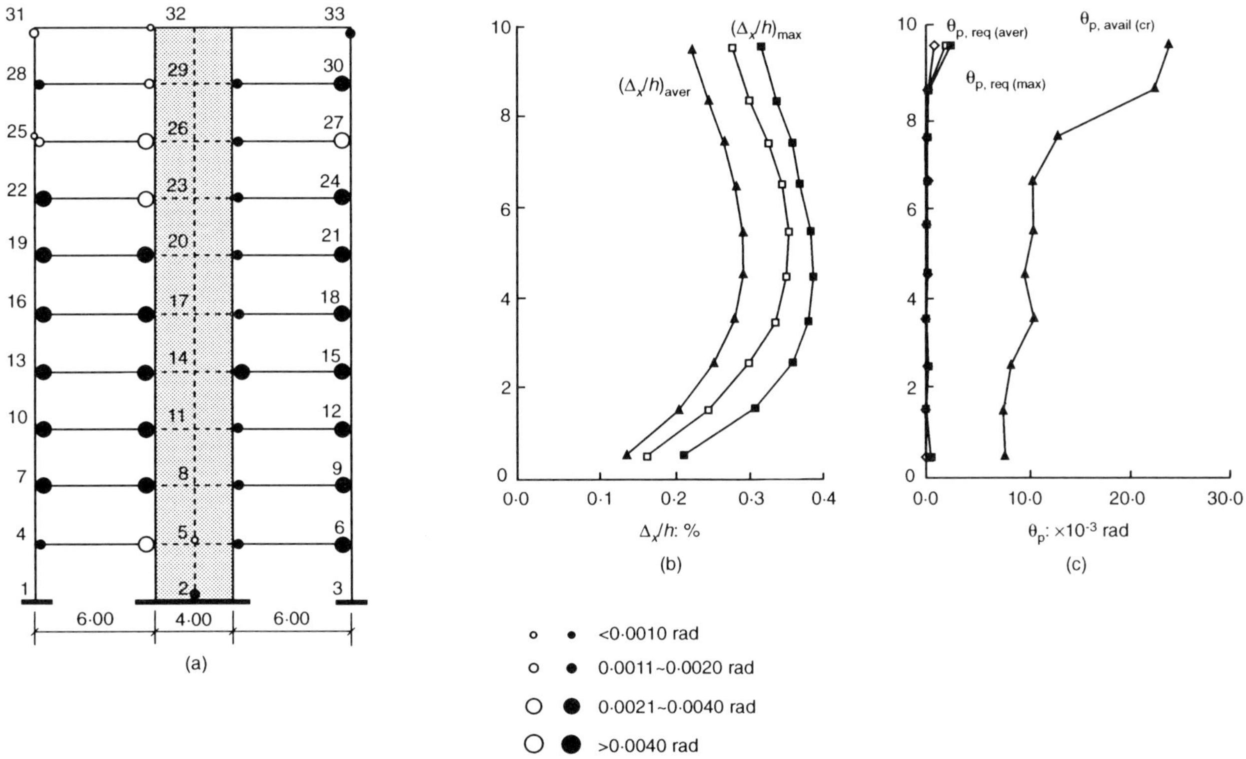

Fig. 5.17. Dual structure designed according to new procedure: (a) plastic hinge distribution for the most critical motion; (b) inter-storey drift ratios; (c) required and available plastic rotations in exterior columns

very little systematic effect on the quantities of concrete and steel. On average, however, DCM has a slight advantage over DCL and DCH for the same design ground motion. Nevertheless, this effect of DC is much smaller than that of the analysis method: in the two cases considered the 'simplified modal response spectrum' or 'equivalent static' analysis leads to about 10% more steel than the 'multi-modal response spectrum' or 'dynamic'. This difference appears mainly in the beams. Finally, regarding the distribution of steel in the structure, it is clear that increasing the ductility class shifts the steel from the beams to the columns: in frames the respective share of beams and columns in the steel total is, on average, about 45–55% for DCL, 40–60% for DCM and 35–65% for DCH. Also, increasing the ductility class transfers some steel from the longitudinal to the transverse direction of the elements: for frames the share of the longitudinal and the transverse directions to the steel total is, on average, 80–20% in DCL, 75–25% in DCM and 60–40% in DCH. Unfortunately, the sample of wall-equivalent dual structures is too small to draw general conclusions for this class of structures as well.

5.3.2.2. Alternative column capacity designs. Figures 5.28 and 5.29 focus on the capacity design magnification of column moments. Fig. 5.28 presents the mean value, over all columns of the storey, of the magnification factor, i.e. of the term $(1 + \gamma_{Rd} - \delta)$ in equation (2.14), for the seismic action combination which produces the maximum (and most likely controlling) column design moment above and below the joint. Fig. 5.29 gives similar results for the case in which the conventional capacity design format, equation (2.12) of section 2.4.3.2, is used, instead of the 'relaxed' one of equation (2.14). These factors are listed separately for each direction

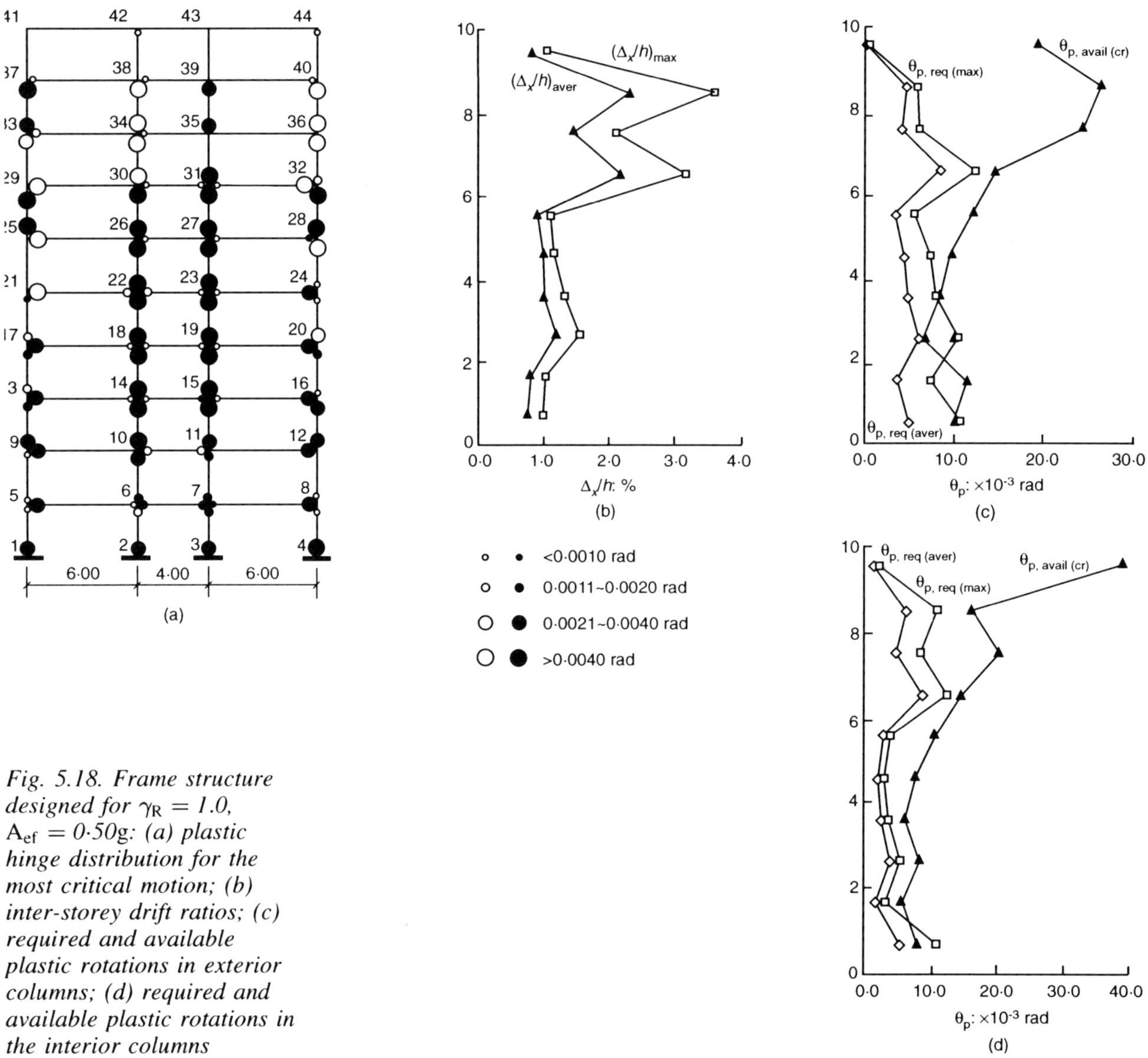

Fig. 5.18. Frame structure designed for $\gamma_R = 1.0$, $A_{ef} = 0{\cdot}50g$: (a) plastic hinge distribution for the most critical motion; (b) inter-storey drift ratios; (c) required and available plastic rotations in exterior columns; (d) required and available plastic rotations in the interior columns

of bending and storey, with the exception of the 12-storey structure in which the results for storeys 6–9 are omitted. The results for the four omitted storeys show, on the average, a gradual increase towards the higher values typically found in the top two storeys. In the 4-, 8- and 12-storey structures the magnification factor for the 2nd-from-the-corner exterior column on the long (X) sides, which is not connected to a beam in the Y direction, is 1·0 in this latter direction. This column is included in the mean, so the average magnification for the other columns is equal to 1·25 times the value in the figures, minus 0·25. The 8-storey core building has only exterior columns, connected to a beam only along the perimeter. So only the corner columns have magnification factors greater than 1·0 in both directions. One-third of the rest have magnification factors of 1·0 in the X direction and the other third do so in the Y direction. As the magnification factors of all columns are included in the average, the mean of the factors which are greater than 1·0 is in this case equal to 1·5 times the value in Figs 5.28 or 5.29 minus 0·5.

The results in Fig. 5.29 demonstrate clearly that the more dominated by gravity loads or by minimum reinforcement the design of the beams, the larger is the magnification factor according to the conventional capacity

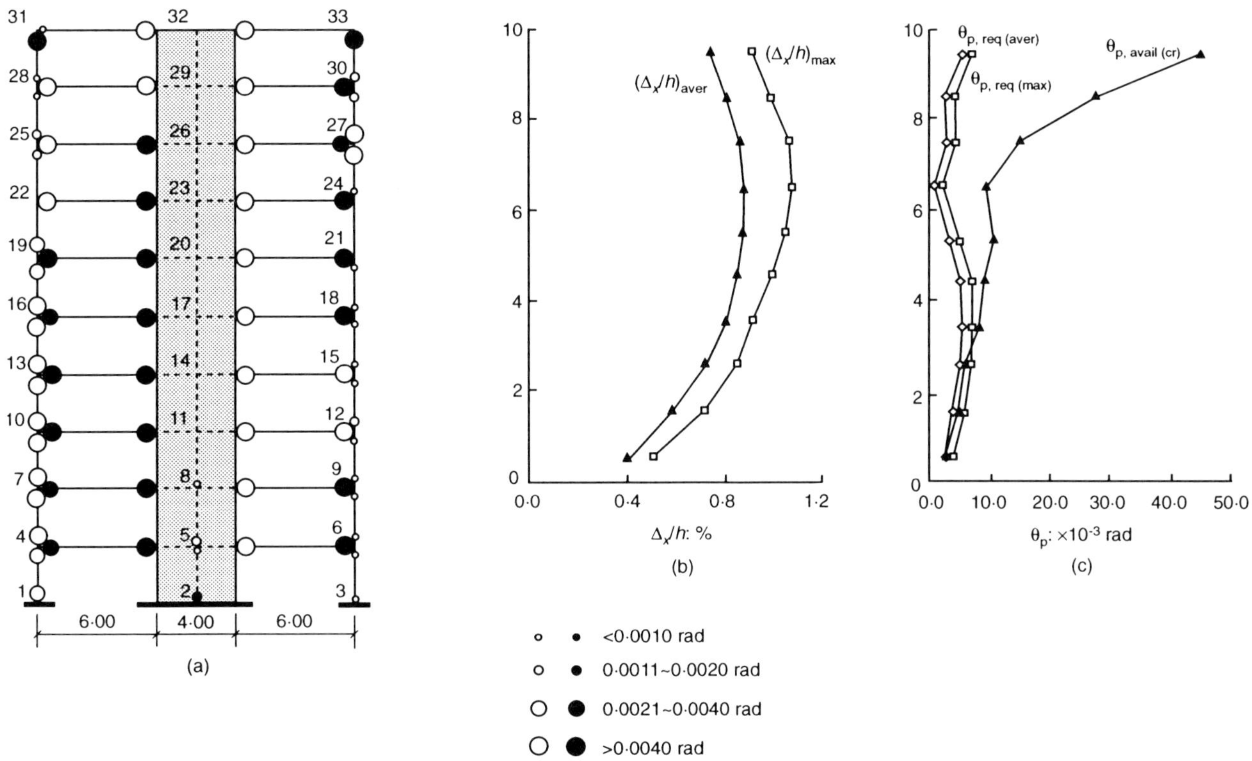

Fig. 5.19. Dual structure designed for $\gamma_R = 1{\cdot}0$, $A_{ef} = 0{\cdot}50g$: (a) plastic hinge distribution for the most critical motion; (b) inter-storey drift ratios; (c) required and available plastic rotations in columns

design format, equation (2.12): this is the case in the top storeys of all the buildings (beams dominated by minimum reinforcement and gravity loads), in the low-rise structures (gravity load dominated beams), in the long-span Y direction of the 3-storey industrial building, and, taking into account the observation at the end of the last paragraph about the real average value, the perimeter columns of the core buildings (gravity load dominated beams). In the 4-, 8- and 12-storey structures the average capacity design magnification is lower in direction Y than in X, not only because of the magnification factor of 1·0 in direction Y of the 2nd-from-the-corner column in direction X, but also because the columns of the two interior frames in direction X are connected to a long-span, gravity load dominated beam. Within the same ductility class (DCM), the average capacity design magnification is higher in the buildings designed for a 0·15 g ground acceleration, as their beams are more dominated by gravity loads than in the DCM 0·3 g structures. The difference in the capacity design magnification between DCM and DCH structures designed for the 0·3 g ground acceleration is small and reflects the difference between the corresponding γ_{Rd} values (1·2 for DCM, 1·35 for DCH). On the contrary, the difference in the magnification between DCL and DCM or DCL and DCH is very large, more than bridging the gap between the elastic seismic moments of the different DCs due to their different q-factors (the ratio of q-factors of DCL to DCM is 1:1·5 and of DCL to DCH 1:2).

The results in Fig. 5.28 show that the relaxation of capacity design according to equation (2.14) is quite effective in reducing the magnification factors. In those cases in which the conventional capacity design magnification factors of Fig. 5.29 are relatively low due to the control of beam strength by the seismic action, the magnification factors in

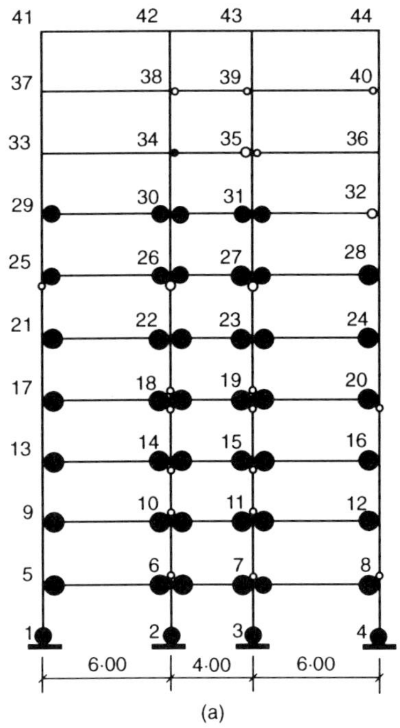

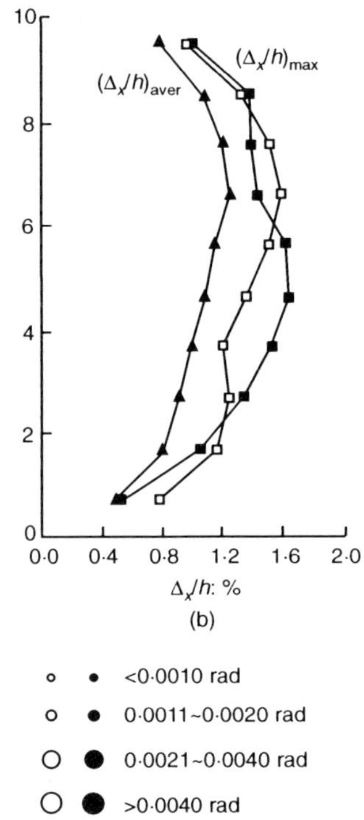

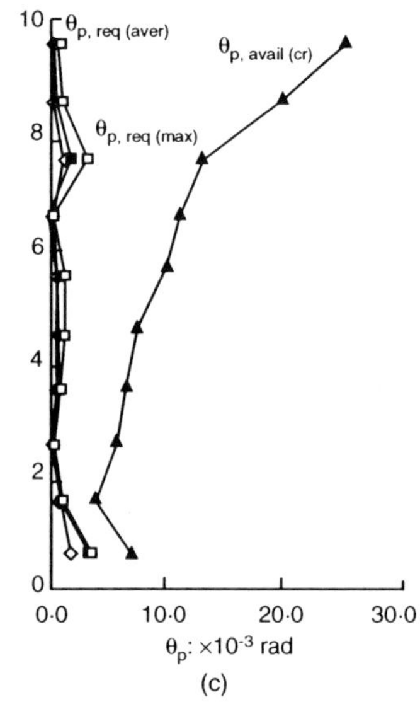

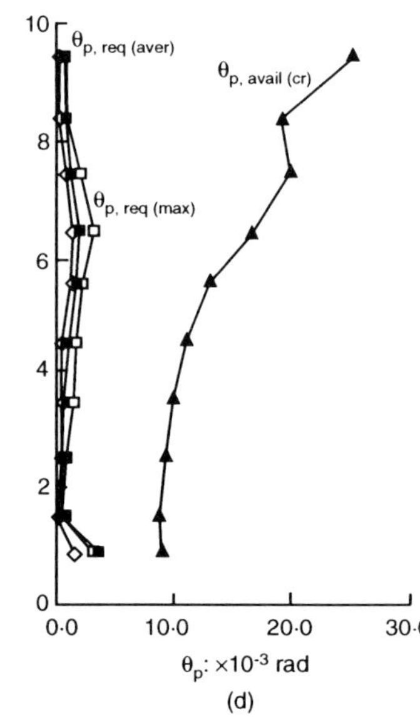

Fig. 5.20. Frame structure designed according to new procedure, $A_{ef} = 0{\cdot}50g$: (a) plastic hinge distribution for the most critical motion; (b) inter-storey drift ratios; (c) required and available plastic rotations in exterior columns; (d) required and available plastic rotations in the interior columns

Fig. 5.29 are only slightly higher than 1·0. In the more gravity load dominated cases, the 'relaxed' capacity design magnification factor is significantly higher than 1·0, but still much lower than the corresponding values in Fig. 5.29. It should be mentioned, however, that the magnification factors in Figs 5.28 and 5.29 are typically produced by different load combinations. Those in Fig. 5.28 correspond to a combination in which the seismic moments in the columns and the beams are relatively large, producing the largest column design moment as the product of a low magnification factor times a large unmagnified column moment M_{Sc}. On the contrary, those in Fig. 5.29 correspond to a load combination in which beam and column moments are relatively low in comparison to the beam strengths M_{Rb}. In this latter case, the largest column design moment is equal to the product of a small column moment from the analysis M_{Sc} times a relatively high capacity design magnification factor. The magnitude of the 'relaxed' capacity design magnification factors still increases with increasing ductility class. However, this magnification factor is not, in general, large enough to counterbalance the difference in column moments from the analysis due to the different q-factors of the ductility classes. So, the final column design

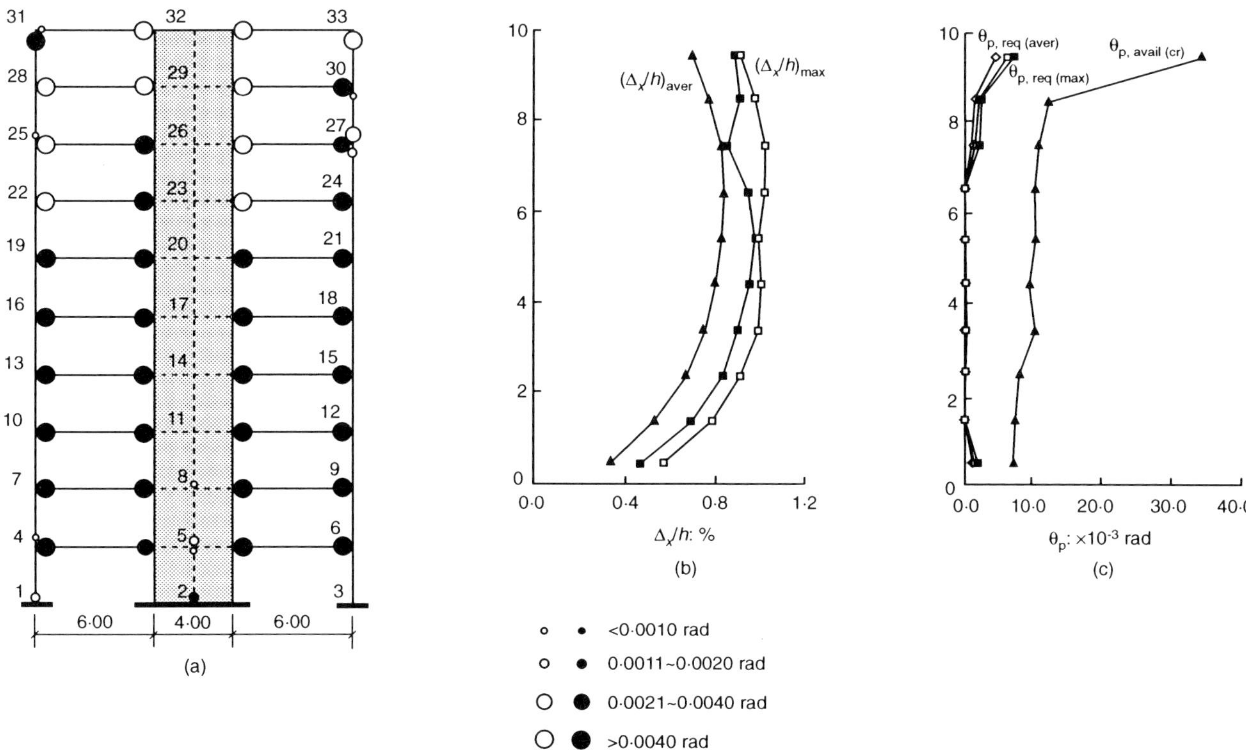

Fig. 5.21. Dual structure designed according to new procedure, $A_{ef} = 0{\cdot}50g$: (a) plastic hinge distribution for the most critical motion; (b) inter-storey drift ratios; (c) required and available plastic rotations in columns

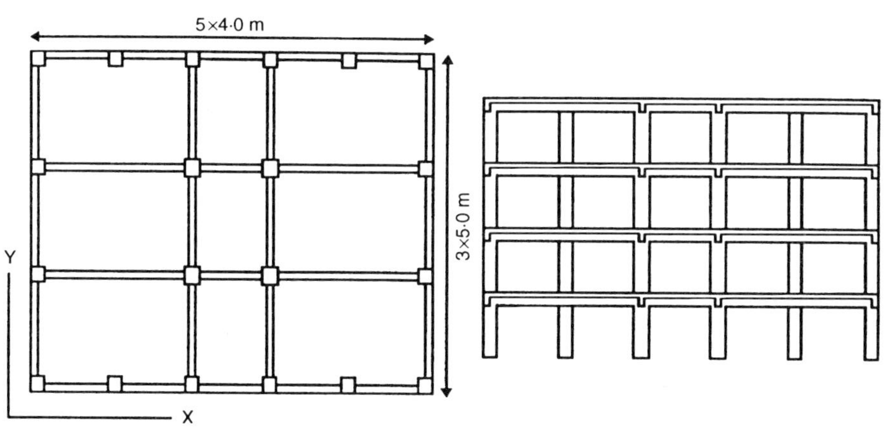

Fig. 5.22. Plan and elevation of 4-storey frame structure

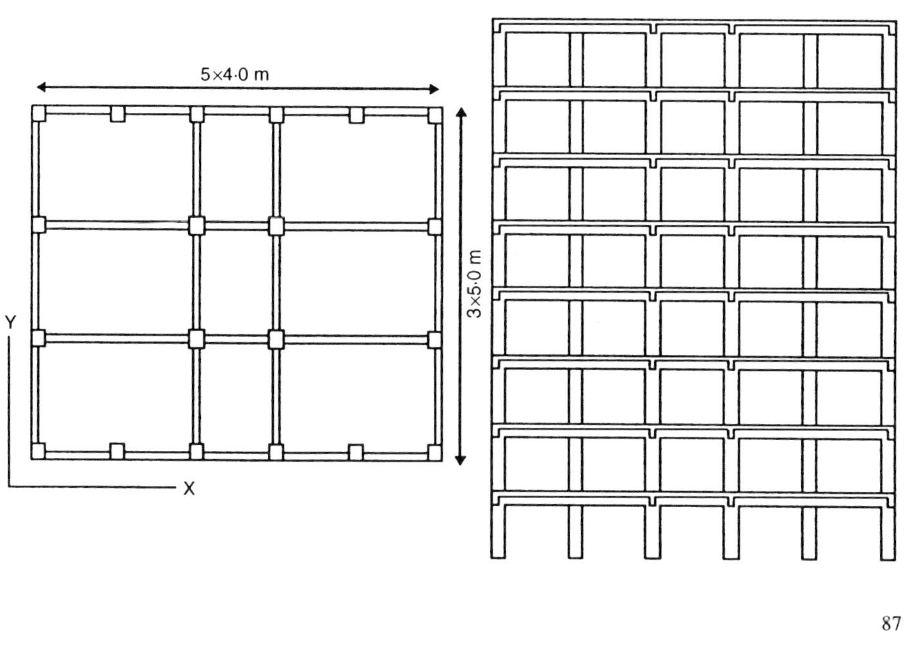

Fig. 5.23. Plan and elevation of 8-storey frame structure

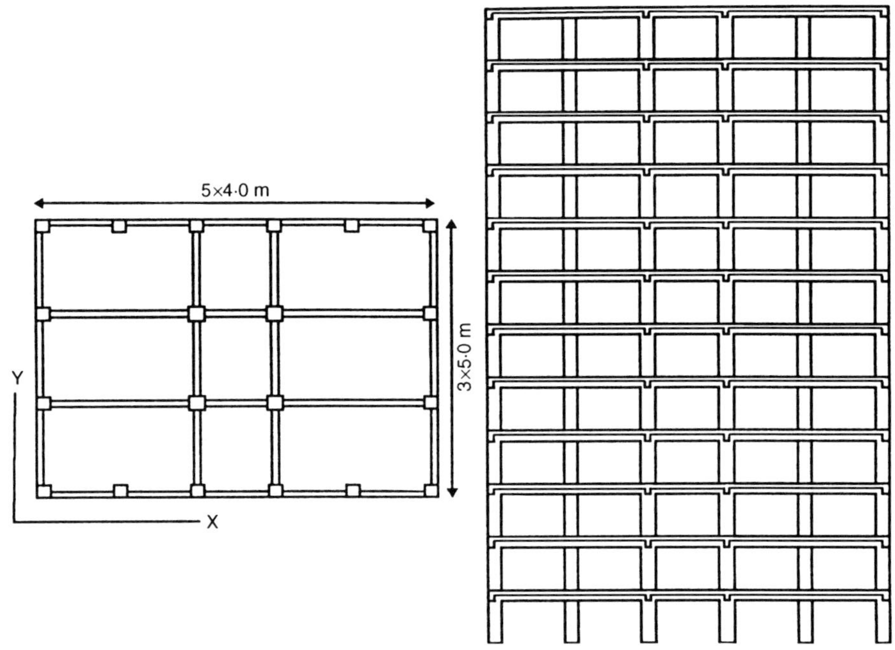

Fig. 5.24. Plan and elevation of 12-storey frame structure

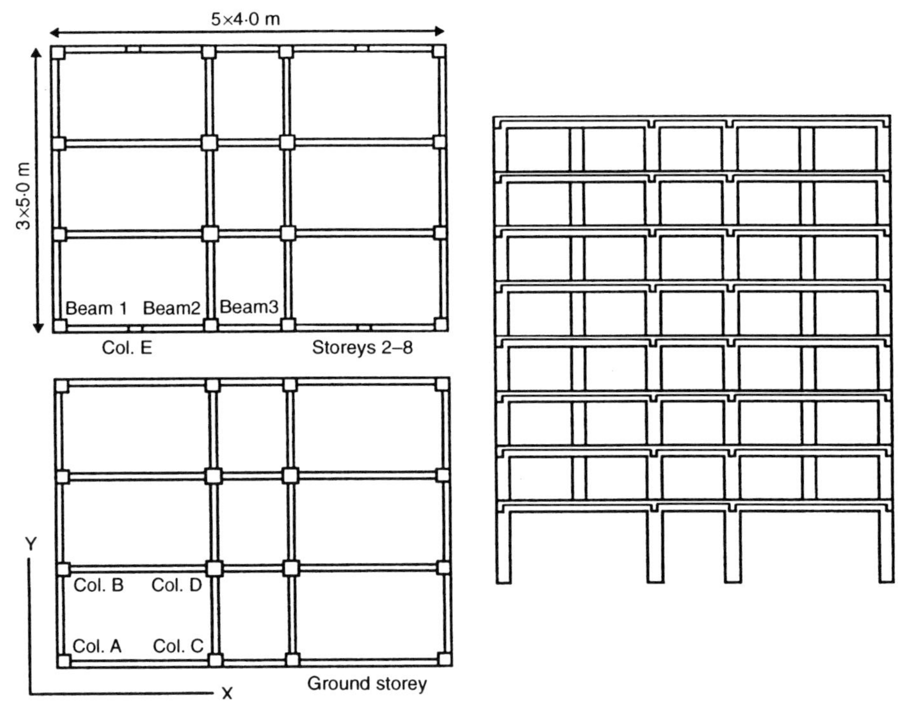

Fig. 5.25. Plan and elevation of 8-storey irregular frame structure

moments, after the capacity design magnification, decrease with increasing DC.

To complete the picture regarding the available overstrengths in the columns, the ratio of the total shear force required for simultaneous development of the design flexural capacities at top and bottom of all columns in a storey $\Sigma M_{Rc}/h$ to the corresponding seismic shear force for elastic response to the design ground motion, is presented in Figs 5.30 and 5.31. Fig. 5.31 refers to columns designed according to the conventional capacity design format, equation (2.12), while Fig. 5.30 applies when capacity design is 'relaxed' according to equation (2.14). For all but the

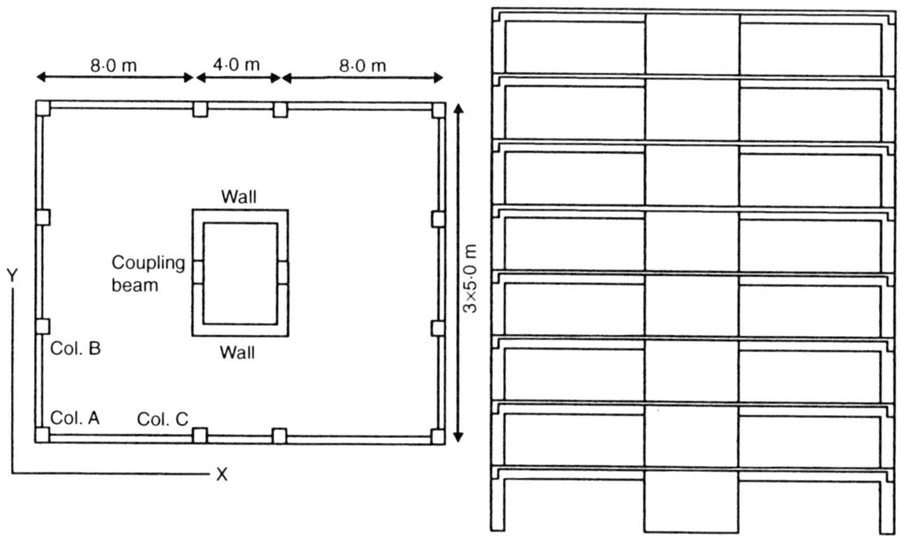

Fig. 5.26. Plan and elevation of 8-storey coupled frame–wall core structure

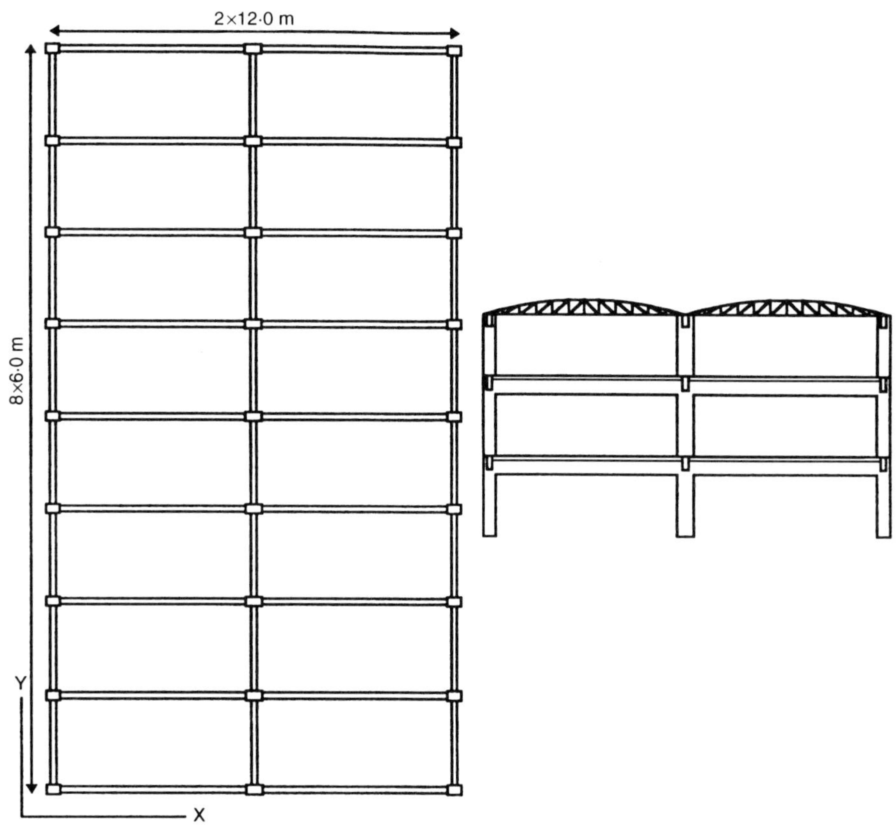

Fig. 5.27. Plan and elevation of 3-storey frame structure

8-storey 'core' buildings, in which the numerator and the denominator of this ratio refers only to the 12 columns on the perimeter of the building, the denominator of the ratio is the total storey shear for elastic response to the design seismic action. Then the value of the ratio is the fraction of the design ground motion at which the storey-sway mechanism is formed at the storey in question, if no other inelastic action has developed elsewhere (i.e. in the beams or in the columns of the other storeys) and if the strength of the materials, steel and concrete, are equal to their design values.

The most noticeable features of the results in Figs 5.30 and 5.31 are the following.

Table 5.2. Global design results for the 26 buildings

Storeys/ config.	a_g/DC	q	T_1 : s		V_b/W: %		Drift ratio: %						Concrete: m^3	Steel				
							Top		Inter-storey		Inter-st./top			Total: t	Beam/ col.: %	Longit./transvers: %		
			X	Y	X	Y	X	Y	X	Y	X	Y				Beams	Columns	Total
4/Fr.	0·15/L	2·50	0·41	0·51	14	14	0·16	0·26	0·22	0·39	1·35	1·30	117	15·3	46/54	83/17	73/27	78/22
	0·30/M	3·75	0·41	0·51	10	10	0·16	0·26	0·22	0·39	1·35	1·30	117	15·5	43/57	73/27	68/32	70/30
	0·30/M	3·75	0·40	0·46	19	19	0·31	0·42	0·40	0·56	1·30	1·35	130	24·7	37/63	78/22	62/38	68/32
	0·30/H	5·00	0·40	0·46	15	15	0·31	0·42	0·40	0·56	1·30	1·35	130	23·0	39/61	66/34	55/45	60/40
8/Fr.	0·15/L	2·50	0·61	0·69	14	13	0·18	0·22	0·26	0·31	1·40	1·40	323	44·5	45/55	90/10	75/25	82/18
	0·15/M	3·75	0·66	0·81	8	9	0·20	0·27	0·28	0·38	1·40	1·40	323	41·4	38/62	75/25	68/32	71/29
	0·30/M	3·75	0·56	0·63	19	20	0·32	0·39	0·44	0·55	1·40	1·40	365	60·4	43/57	83/17	74/26	78/22
	0·30/H	5·00	0·56	0·63	15	16	0·32	0·39	0·44	0·55	1·40	1·40	365	68·2	35/65	71/29	58/42	62/38
8/Irr.	0·15/L	2·00	0·78	0·89	12	12	0·19	0·21	0·26	0·29	1·35	1·40	362	54·6	43/57	88/12	76/24	81/19
	0·15/M	3·00	0·78	0·89	8	8	0·19	0·21	0·26	0·29	1·35	1·40	362	52·7	39/61	77/23	71/29	73/27
	0·30/M	3·00	0·76	0·87	16	16	0·32	0·36	0·44	0·51	1·35	1·40	414	80·1	40/60	84/16	74/26	78/22
	0·3/M-St	3·00	0·76	0·87	20	20	0·39	0·43	0·52	0·60	1·35	1·35	414	86·7	43/57	85/15	74/26	79/21
	0·30/H	4·00	0·73	0·82	12	12	0·32	0·36	0·44	0·51	1·35	1·40	414	81·9	35/65	74/26	59/41	64/36
12/Fr.	0·15/L	2·50	0·66	0·72	17	16	0·17	0·21	0·24	0·29	1·40	1·40	643	97·2	37/63	89/11	76/24	80/20
	0·15/M	3·75	0·66	0·72	11	11	0·17	0·21	0·24	0·29	1·40	1·40	643	94·4	32/68	78/21	72/28	74/26
	0·30/M	3·75	0·58	0·63	23	23	0·34	0·40	0·47	0·56	1·40	1·40	682	117	42/58	84/16	72/28	78/22
	0·3/M-St	3·75	0·58	0·63	25	25	0·42	0·51	0·58	0·70	1·35	1·40	682	130	46/54	85/15	72/28	78/22
	0·30/H	5·00	0·58	0·63	18	18	0·32	0·37	0·43	0·52	1·35	1·40	712	126	36/64	74/26	58/42	64/36
3/Fr.	0·15/L	2·50	0·74	0·56	13	14	0·36	0·25	0·45	0·28	1·25	1·10	428	47·3	50/50	84/16	76/24	80/20
	0·15/M	3·75	0·74	0·56	9	10	0·36	0·25	0·45	0·28	1·25	1·10	428	50·9	46/54	79/21	74/26	76/24
	0·30/M	3·75	0·51	0·40	19	19	0·40	0·26	0·49	0·30	1·20	1·15	520	72·7	37/63	81/19	77/23	78/22
	0·30/H	5·00	0·51	0·40	15	14	0·40	0·26	0·49	0·30	1·20	1·15	520	79·1	33/67	72/28	60/40	64/36
8/cor.	0·15/L	1·75	0·58	0·53	18	19	0·19	0·15	0·24	0·18	1·30	1·20	224	44·5	24/76	86/14	48/52	57/43
	0·15/M	2·62	0·58	0·53	12	12	0·19	0·15	0·24	0·18	1·30	1·20	224	38·6	23/77	76/24	42/58	50/50
	0·30/M	2·62	0·50	0·47	25	26	0·28	0·23	0·35	0·28	1·25	1·20	310	68·5	25/75	85/15	49/51	58/42
	0·30/H	3·50	0·50	0·47	19	20	0·28	0·23	0·35	0·28	1·25	1·20	310	65·7	22/78	76/24	42/58	50/50

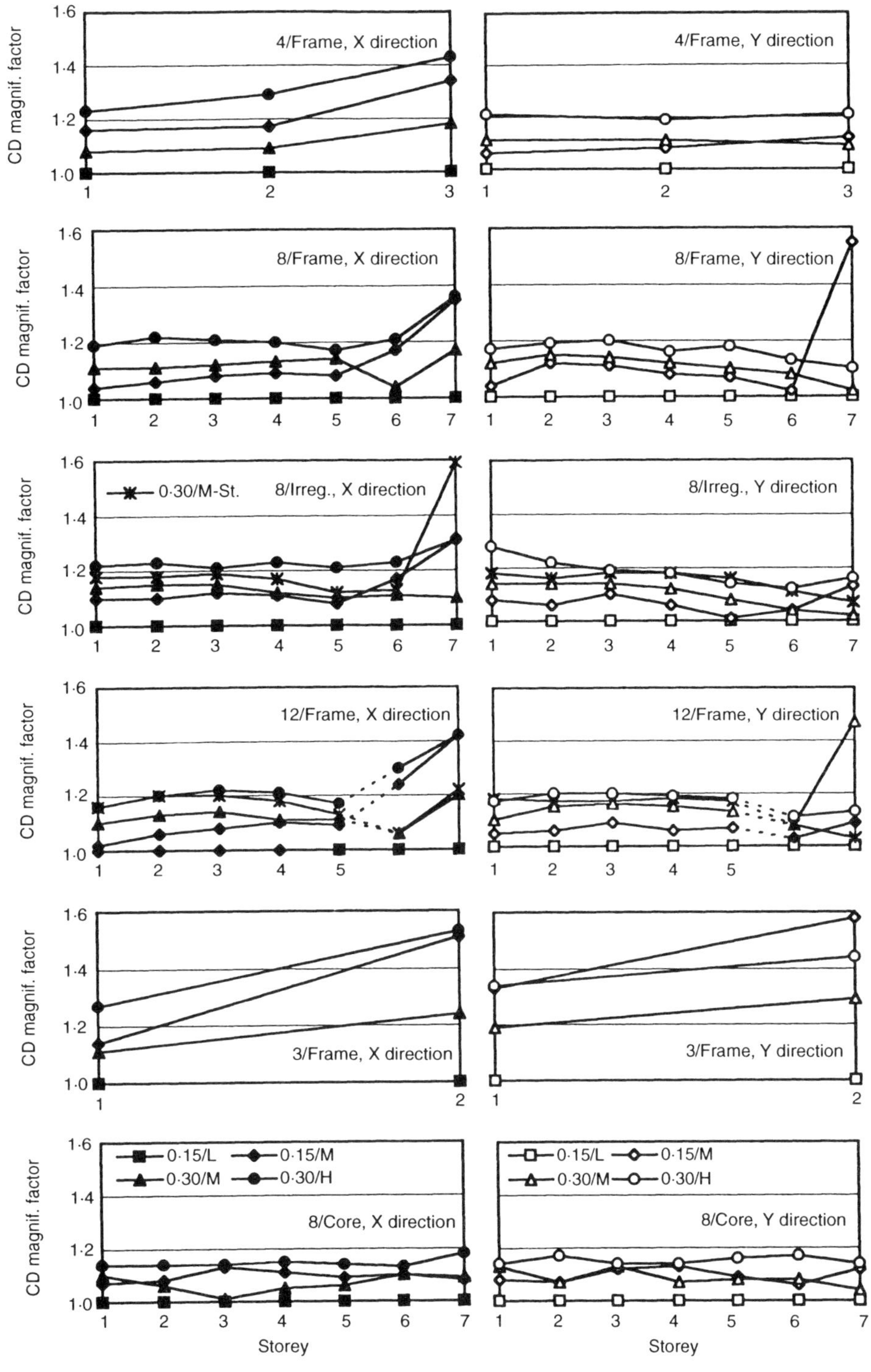

Fig. 5.28. Storey–average capacity design moment magnification factors for columns of 26 buildings with EC8 relaxation due to beam moment reversal, equation (2.14)

(*a*) When the Eurocode 8 approach of 'relaxing' capacity design is followed, the ductility class has little effect on column capacities, whereas the conventional capacity design leads to column capacities which are higher in DCM than in either DCL or DCH structures of the same configuration and design ground acceleration.

(*b*) The values in these figures arc, in general, of the order of 1·0 (strictly speaking they are, on average, around 1·0 in the lower storeys of the structures designed for a ground acceleration of 0·15 g, or slightly below 1·0 in those of the 0·30 g structures, and increase above 1·0 in the upper storeys).

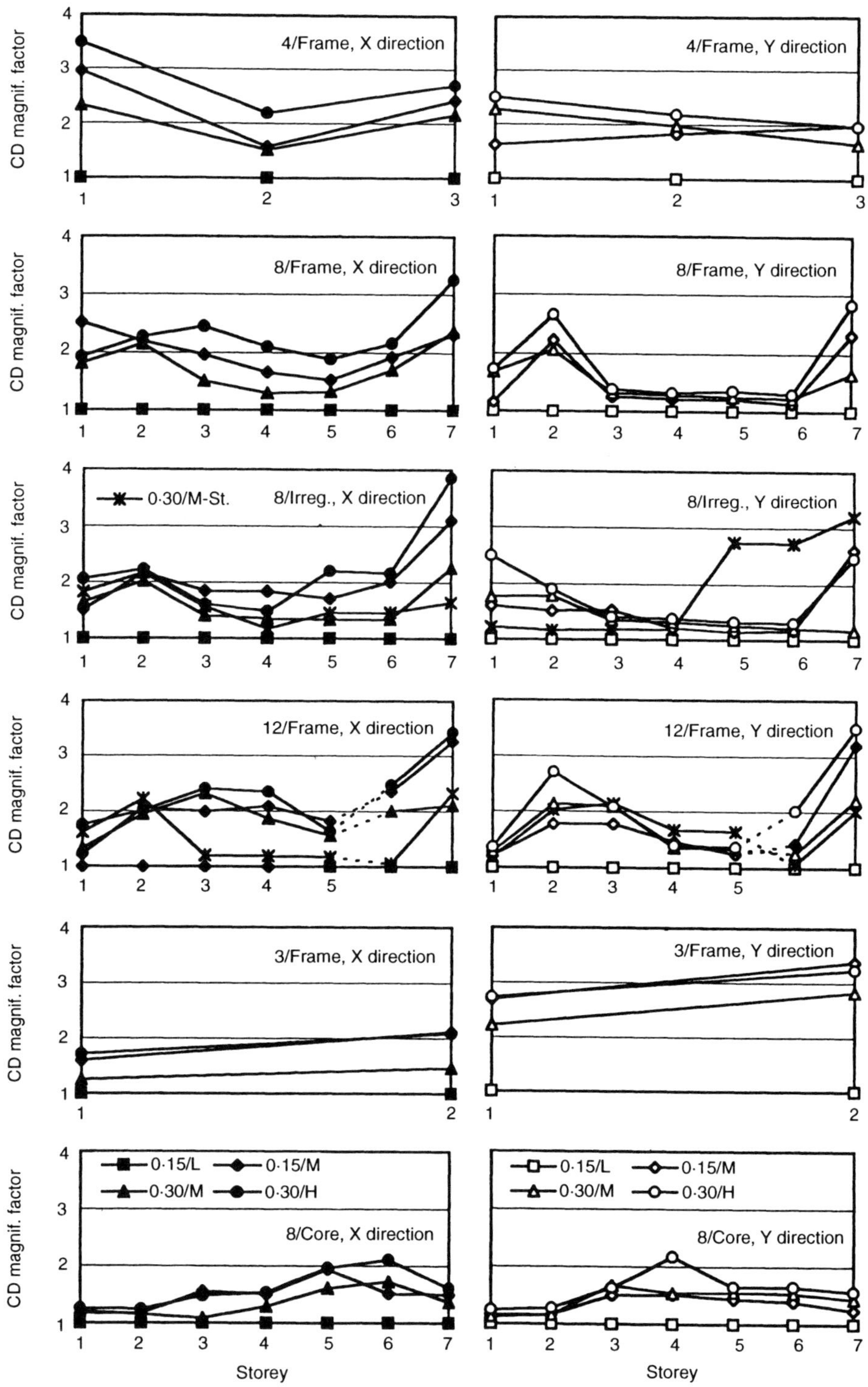

Fig. 5.29. Storey–average capacity design moment magnification factors for columns of 26 buildings with the conventional capacity design format, equation (2.12)

(*c*) The large differences in the moment magnification factor between the two alternative capacity design strategies, the conventional one of equation (2.12) and the 'relaxed' of equation (2.14) allowed by Eurocode 8 (cf. Figs 5.28 and 5.29), do not lead in general to similar differences in column flexural capacities.

One reason for point (*a*) is that most columns, especially in the upper half of the building, have the minimum reinforcement ratio, which is equal to 1% independent of the DC. Another reason is that the capacity design moment magnification factors in Figs 5.28 and 5.29, combined with the

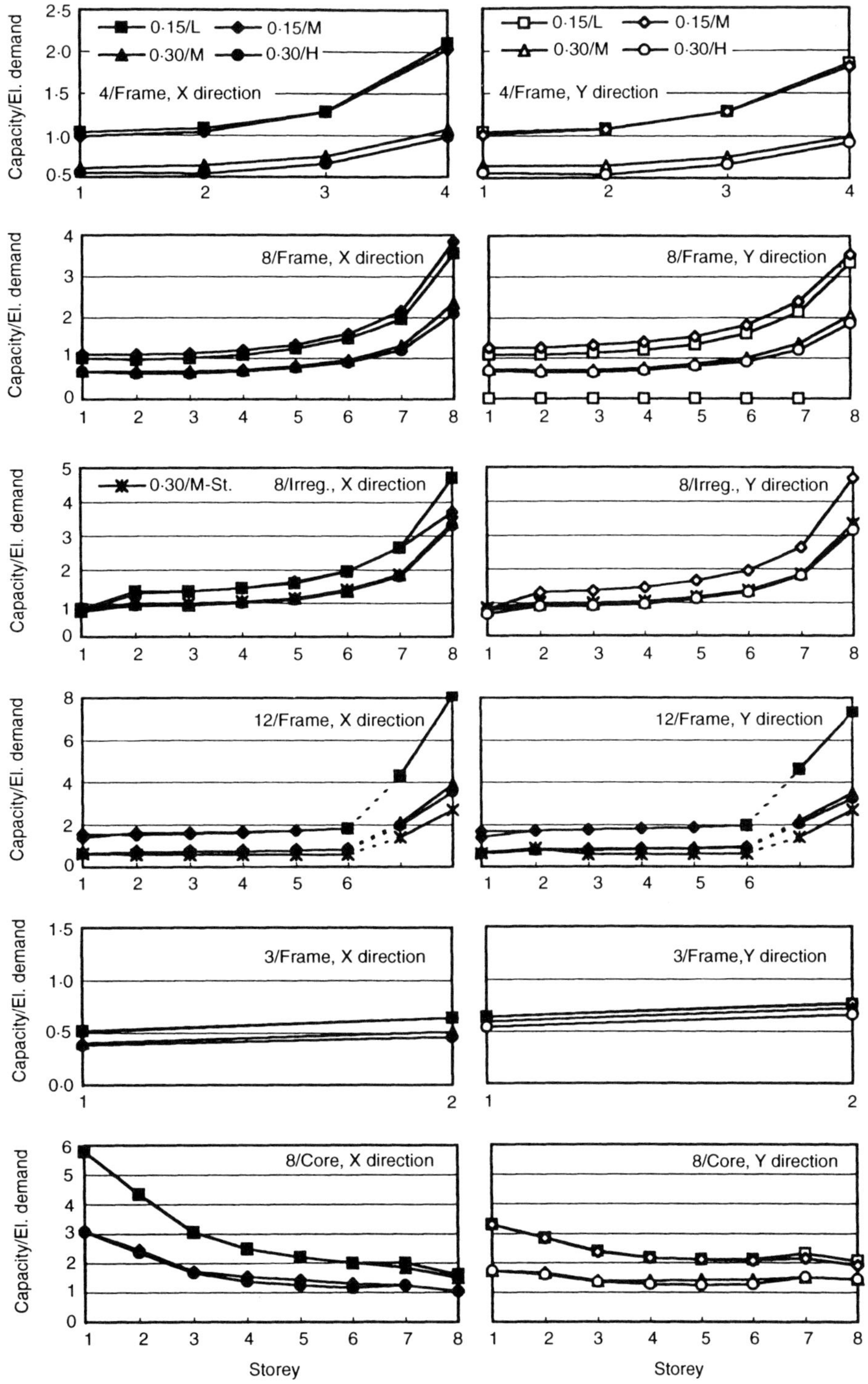

Fig. 5.30. Sum of column top and bottom capacities in a storey divided by the storey elastic seismic demand for "relaxed" capacity design (Fig. 5.28, equation (2.14))

larger number of column intermediate bars required in the higher ductility classes, are enough to bridge the difference in the q-factors between the three DCs, or even surpass it for DCM and DCL structures, if the conventional capacity design format is applied to the former. Observation (*b*), i.e. the fact that most values in the figures are around 1·0 and higher, is due to an accumulation of overstrengths in the columns. In addition to the overstrengths mentioned above, i.e. the capacity design magnification for DCM and DCH and the detailing rules for vertical reinforcement, there is a systematic overstrength in one of the top or bottom sections due to the fact

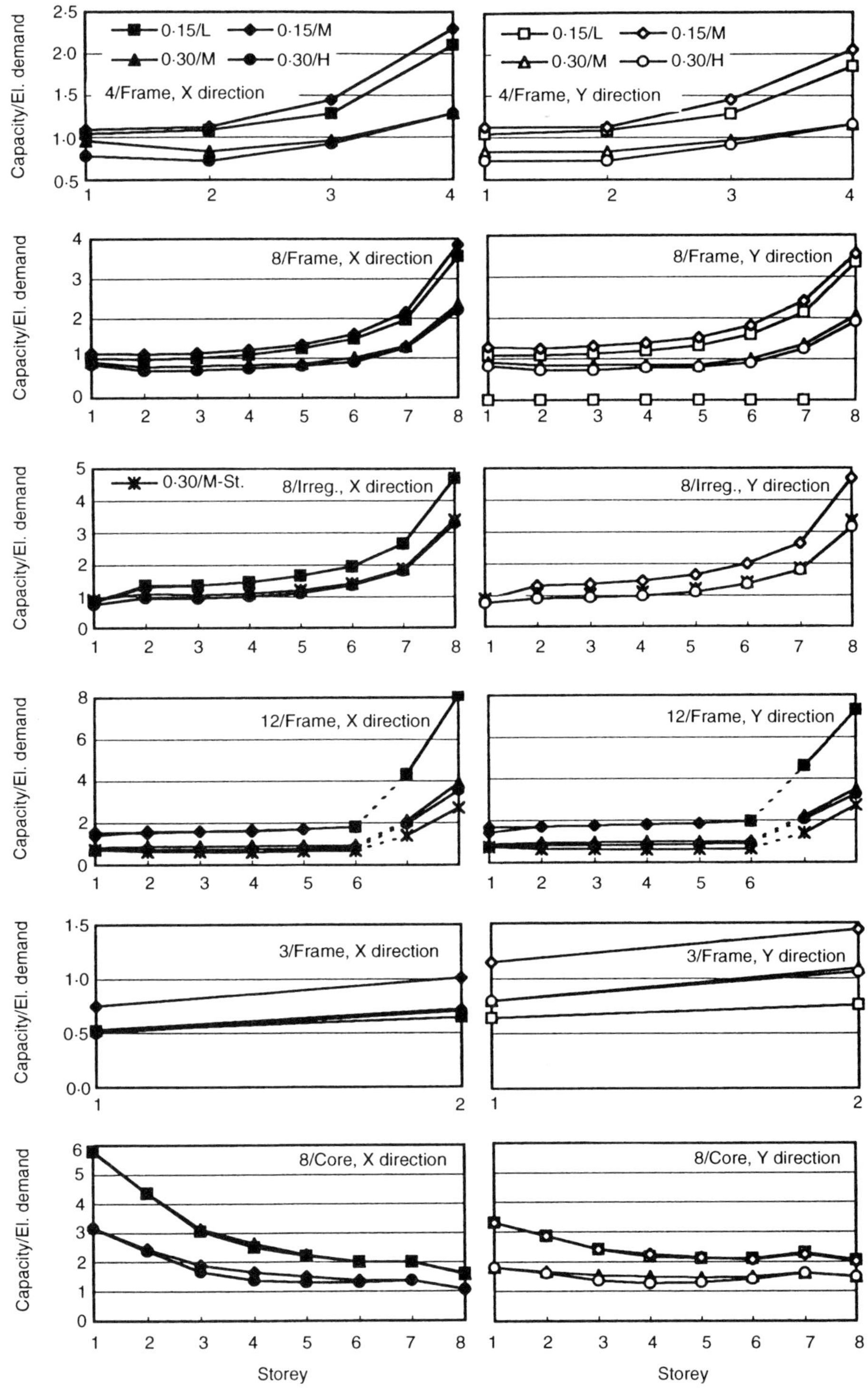

Fig. 5.31. Sum of column top and bottom capacities in a storey divided by the storey elastic seismic demand for conventional capacity design (Fig. 5.29, equation (2.12))

that column sections above and below a joint have the same reinforcement, although their elastic moments M_{Sc} are generally very different. Moreover, there is an overstrength due to the fact that columns have been designed for action effects which are not reflected in the denominator of the ratio, i.e. in the storey seismic shear. They have been designed for the effects of an accidental eccentricity of the horizontal seismic action, for biaxial bending due to a simultaneous orthogonal component equal to 30% of the design seismic action, and (especially in the upper storeys) for simultaneously acting biaxial moments due to the quasi-permanent gravity loads. Because the storey seismic shear decreases in the upper storeys, whereas the overstrengths due to minimum reinforcement or gravity load effects do not, the ratios in Figs 5.30 and 5.31 increase from the ground to the top. This is not the case in the 8-storey core building, in which the fraction of the storey shear taken by the frames increases from nearly zero at the ground storey to about 100% at the top. Finally, it is mainly the column overstrengths due to minimum reinforcement and detailing requirements that moderate the effect of the large difference in the moment magnification factors resulting from the two alternative capacity design approaches, those of equations (2.12) and (2.14). It is noteworthy that although the relaxation of capacity design allowed by Eurocode 8 according to equation (2.14) leads to columns of lower strength in comparison to the conventional capacity design procedure of equation (2.12), it produces much more uniform column capacities among the alternative DCs.

The lower values given in Figs 5.30 and 5.31 for the two DCM structures designed using the results of the 'simplified modal' (i.e. 'equivalent static') analysis, in comparison to those designed on the basis of the 'multi-modal response spectrum' ('dynamic') analysis method, are due to the different storey shears of the two approaches: storey shears are significantly higher when the 'equivalent static' approach is used (by 7–25% in the 8-storey irregular building and by 20–50% in the 12-storey one, with the larger differences appearing in the top storeys). So, although the 'equivalent static' analysis leads to stronger columns in absolute terms, the strengths appear lower when normalized to the design storey shear. It is also noteworthy that, despite the application of Eurocode 8 provisions accounting for irregularities in elevation, the effects of such irregularities of the 8-storey irregular structures are still evident in the column capacity results in Figs 5.30 and 5.31.

Similar calculations for the beams have shown that, with the exception of the upper storeys, in which minimum reinforcement and gravity loads may exert control, the as-designed beams (i.e. essentially neglecting participation of slab reinforcement to the tension flange) possess very little or no overstrength for the seismic action: in most beams the combination of the gravity action effects with the design seismic moments from the analysis, positive or negative, is enough for the design flexural capacity to be reached. So, for beams, the ratios of the type shown in Figs 5.30 and 5.31, but considering also the simultaneous bending due to gravity loads, are of the order of $1/q$ for negative bending (tension at the top), or slightly higher for positive, due to the presence of bottom reinforcement at least equal to 50% of that at the top. This lack of beam overstrength, combined with the significant overstrength of the columns against soft-storey formation, is in full accordance with the strong column–weak beam capacity design philosophy of modern seismic standards. However, significant overstrength may be introduced in the beams by the (much larger than considered in design) participation of slab reinforcement, adversely modifying the desired column–beam strength ratio.

In section 5.3.4, the inelastic response of the core buildings in the horizontal direction Y (in which the walls are coupled) and of the 12-storey and the irregular frame buildings in direction X, in which the irregularity in elevation is more prominent, is computed and discussed. In section 5.3.5 the inelastic seismic response of the 4-storey, the 12-storey and the 3-storey frame structures is studied. In both cases input ground motions at and above the design motion intensity are applied, as described in section 5.3.3. In section 5.3.4 only the set of structures with columns proportioned according to the relaxed capacity design of equation (2.14) are considered, as representing the standard approach of Eurocode 8. Section 5.3.5 considers

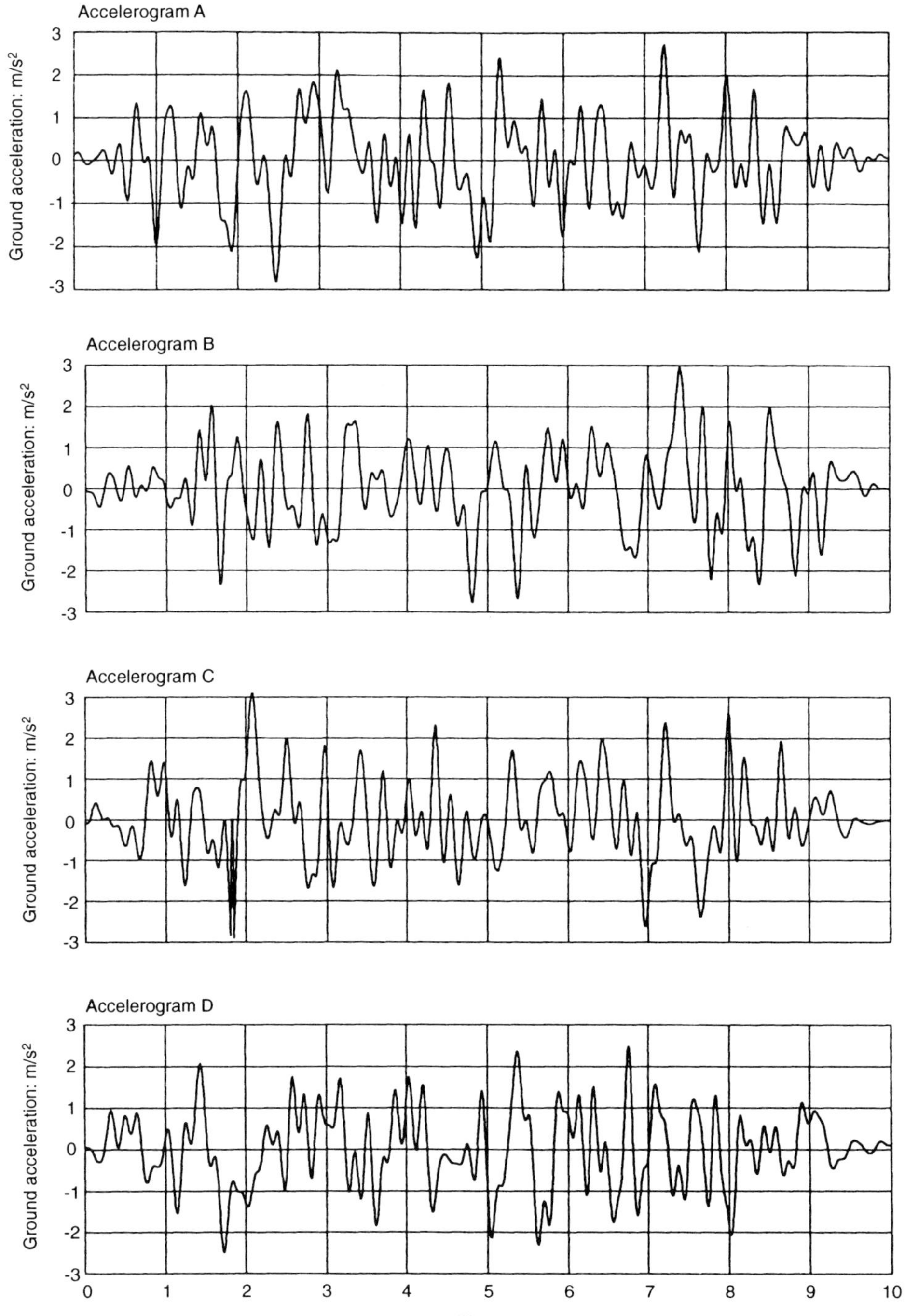

Fig. 5.32. Time–histories of artificial accelograms, normalized to 0·3 g effective peak acceleration

both sets of 4-storey, 12-storey and 3-storey structures, i.e. those with columns proportioned according to equation (2.14) or to equation (2.12) of section 2.4.3.2.

5.3.3. Seismic input motions for the inelastic response analysis

The non-linear dynamic response analyses in sections 5.3.4 and 5.3.5 are performed using as an input four motions artificially generated to conform with the Eurocode 8 elastic response spectrum for 5% damping and Subsoil Class B, for which the structures were designed. The acceleration records, shown in Fig. 5.32, have a duration of 10 s, which includes relatively short rise and fall times. The motions were generated through a variant of

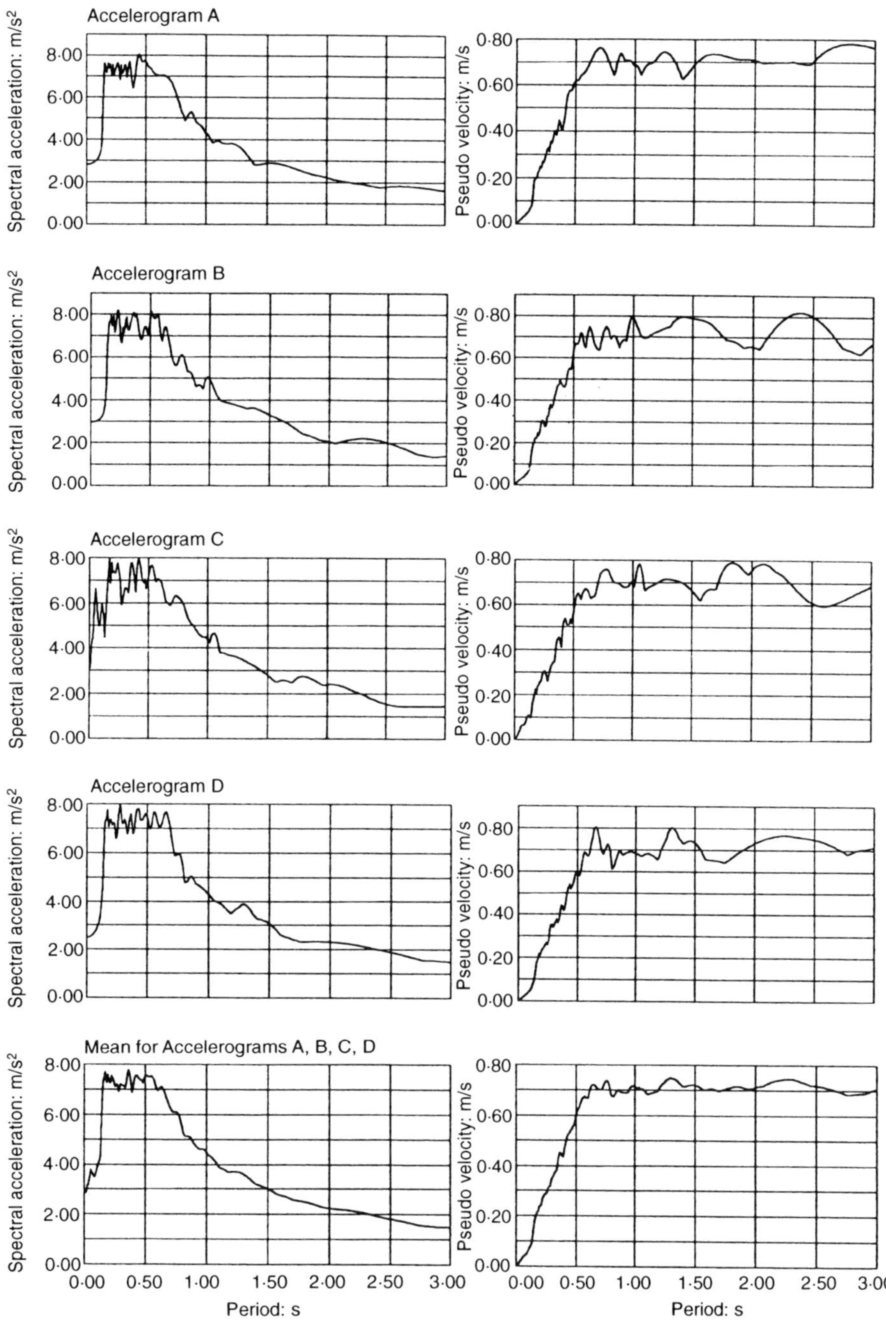

Fig. 5.33. Five per cent damped elastic acceleration and pseudovelocity spectrum for four artificial motions (individual and average spectra)

Vanmarke's SIMQUE approach, which guarantees almost full conformance to the target spectrum. The 5% damped acceleration and velocity elastic spectra of the four motions are shown in Fig. 5.33.

5.3.4. Effect of design and detailing criteria on the response of frame and dual structures

5.3.4.1. Analytical model and method. Three frame configurations from the set of structures described in 5.3.2 above are studied herein in more detail. These are

(*a*) coupled frame–core 8 storeys (Core),
(*b*) irregular 8-storey frame (Irr) and
(*c*) regular 12-storey frame (Reg).

Reference is made to ductility class (DC) low (L), medium (M) or high (H) and design acceleration 0·15 g (015) or 0·3 g (030). Plans and elevations for the buildings studied hereafter are shown in Figs 5.26, 5.25 and 5.24 of section 5.3.2 above, respectively. The frames are discretized using plane frame elements for beams and columns with no special provisions for beam–column connections. The choice of two-dimensional planar modelling is justified herein in the light of the satisfaction of code regularity criteria in the out-of-plane direction and savings in terms of computer analysis time. This is especially true for behaviour factor calculations where inelastic dynamic analysis is repeated to pinpoint the yield and the ultimate limit states. For the three configurations considered, combination of the interior and perimeter lateral load resisting systems is achieved by means of an 'overlay' approach, whereby both are combined in one plane and modelled to have common finite element nodes.

The program used in this study is the adaptive non-linear dynamic analysis package ADAPTIC, the capabilities of which are described elsewhere (Izzuddin and Elnashai, 1989). Frame elements with cubic shape functions accounting for the spread of inelasticity across the section as well as the member length were employed. Material models for concrete, accounting for active confinement and reinforcing steel with kinematic hardening (Elnashai and Izzuddin, 1993, Madas and Elnashai, 1992, Martinez-Rueda and Elnashai, 1995) characterize the response. The concrete model operates on the fibre level and uses loading and unloading branches with stiffness degradation commensurate with the level of straining experienced at any given instance. For steel, a bilinear hardening model is used, since it was not deemed necessary to employ the more complex and computationally taxing multi-surface plasticity model available in ADAPTIC.

Large displacement effects are accounted for by deriving element matrices in a convected axes set using an Eulerian formulation, hence P–δ effects are accurately included. The Hilber–Hughes–Taylor α-integration scheme was implemented for the solution of the equations of motion. This numerically dissipative algorithm, for which accuracy is unaffected by the time-step length, allows effective response predictions to be obtained for highly inelastic structures (Broderick *et al.*, 1994). ADAPTIC has been verified by comparison with closed-form solutions for large displacements and elasto-plastic response as well as experimental results for steel, composite and RC structures.

The above features of ADAPTIC, coupled with an extremely powerful solver and solution strategies (automatic adjustment of load–displacement increment, automatic adjustment of integration time-step, retaking of non-convergent steps from the last equilibrium position and automatic switching from one iterative scheme to another) enabled all analyses to proceed without incident up to the attainment of structural collapse limit states. For the analysis of the twelve structures under the design earthquake, twice design, yield and ultimate, a total number of analyses in excess of 300 was undertaken.

5.3.4.2. Assessment criteria. The seismic response of the structures is assessed by investigating the response under the design earthquake, twice the design acceleration, at yield and at the attainment of an ultimate limit state. With regard to the response under the design earthquake or its multiple, the response parameters studied are the total shear force, member curvature ductility demand (ratio of maximum observed curvature and yield curvature) and inter-storey and global displacement. Two yield and two ultimate limit states have been observed throughout. The two yield criteria are

(*a*) yield of main tensile reinforcement in any member
(*b*) global yield displacement obtained from static analysis assuming an equivalent elasto-plastic system.

Whereas it is recognized that a larger number of local and global ultimate limit state criteria should in general be used, two criteria were selected for this study. This is justified by noting that it is rare, although not impossible, that other criteria will govern (such as a shear criterion employed for the low ductility frame–core structure). The two criteria employed herein, after careful consideration, are

(i) attainment of a critical concrete strain in the confined concrete core
(ii) a 3% inter-storey drift limit at any level.

Both of the ultimate limit state criteria are controversial. To substantiate the former, a study was conducted on three expressions for confined concrete strain from the literature, which lead to the criterion adopted of a 15% drop in post-peak confined stress (this is different to the EC8 definition, which links the ultimate strain to the unconfined concrete strain). As for the drift limit adopted, it is recognized that recent tests have indicated that RC frames may undergo drifts significantly larger than 3% prior to collapse. However, the 3% limit provides a safe lower bound and also facilitates comparisons with earlier studies, the majority of which utilize this value. The definitions of yield and ultimate limit states enable the evaluation of response modification factors (or behaviour factors q). This is discussed further below.

For a particular response period, the structural behaviour factor q, as employed in Eurocode 8, is the ratio of the ordinates of the elastic acceleration spectrum used to define the seismic hazard of a site to those of the inelastic spectrum employed in the derivation of the seismic design forces. Thus

$$q = (S_{\mathrm{a}})_{\mathrm{d}}^{\mathrm{el}} / (S_{\mathrm{a}})_{\mathrm{d}}^{\mathrm{in}} \tag{5.1}$$

where, as shown in Fig. 5.34, $(S_{\mathrm{a}})_{\mathrm{d}}$ is the design spectral acceleration and the superscripts 'el' and 'in' refer to its elastic and inelastic values, respectively.

In EC8, maximum allowable q-factor values are specified for a range of structural forms and construction materials. These values reflect the ability

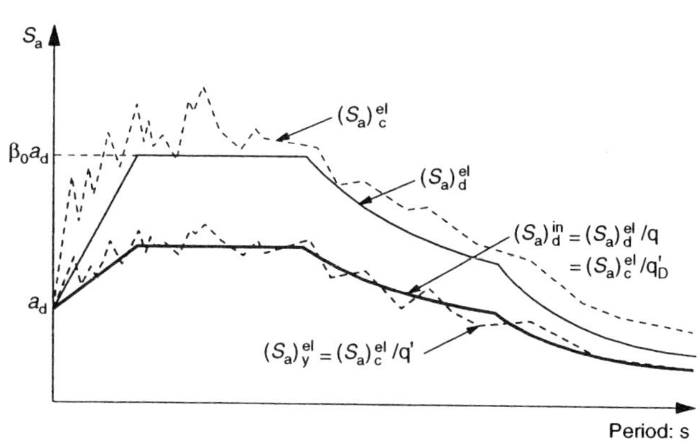

Fig. 5.34. Definition of behaviour factors q and q'_{D}

of each type of structure to undergo stable oscillations in the inelastic range. They are also intended to represent lower bounds on the actual ductility supply of individual structures, thus it should be expected that $q_{\mathrm{code}} < q'$, where q' is the ultimate value which relates the response spectra corresponding to the ground motion intensities which produce structural collapse and yield, viz

$$q' = \frac{(S_{\mathrm{a}})_{\mathrm{c}}^{\mathrm{el}}}{(S_{\mathrm{a}})_{\mathrm{y}}^{\mathrm{el}}} \tag{5.2}$$

where the subscripts 'c' and 'y' denote collapse and yield respectively.

Comparison of equations 5.1 and 5.2 shows that the design code behaviour factors correspond to the case where collapse is imminent under the loading defined by the elastic design spectrum and yield is equally imminent with the conditions of the inelastic design spectrum. This definition of behaviour factor does not account for the effect of overstrength at the yield level (where the overstrength of the structure increases, it is likely that q' will decrease). Fig. 5.34 illustrates the actual ground motion dependence of these conditions. For a given ground motion, however, the collapse intensity will be uniquely defined; in which case by combining equations 5.1 and 5.2 it is possible to further define the behaviour factor of an individual structure as

$$q'_{\mathrm{D}} = (S_{\mathrm{a}})_{\mathrm{c}}^{\mathrm{el}} / (S_{\mathrm{a}})_{\mathrm{d}}^{\mathrm{in}} \tag{5.3}$$

This definition has been employed by Kappos (1991) in the investigation of the behaviour of reinforced concrete buildings. While it has the advantage that it relates the intensity of loading at collapse to the design loads, it does not take into account the disparity between the ground acceleration at yield — which will vary with ground motion, structural design and response period — and that related to the inelastic design spectrum. However, it exhibits the advantage of reflecting the level of overstrength inherent in the design.

By assuming a constant dynamic acceleration amplification, the ratios in equations 5.2 and 5.3 can be represented by the peak ground accelerations corresponding to each of the spectra of Fig. 5.34. Thus

$$q' = \frac{a_{\mathrm{g(collapse)}}}{a_{\mathrm{g(yield)}}} \tag{5.4}$$

$$q'_{\mathrm{D}} = \frac{a_{\mathrm{g(collapse)}}}{a_{\mathrm{g(design)}}/q} \tag{5.5}$$

where $a_{\mathrm{g(collapse)}}$ and $a_{\mathrm{g(yield)}}$ are the peak ground accelerations at which collapse and yield occur respectively, and $a_{\mathrm{g(design)}}$ is the design value of the ground acceleration, denoted as a_{d} in Fig. 5.34.

The definition of equation 5.4 may be regarded as the inherent behaviour factor for a structure the design of which is of an unknown provenance. The problem with this definition is that if q is known and the value resulting from this definition is lower, the implication would be that a lower behaviour factor should have been used in design. If, however, the design is repeated with a lower behaviour factor, it is likely that the resulting analytical behaviour factor will drop down even further. This is because for most structures the yield limit state is usually more sensitive to increases in reinforcement ratios than the ultimate limit state. Therefore, definition 5.4 is not suitable for validating the design behaviour factor. On the contrary, the definition given by 5.5, which may be viewed as the design-implied behaviour factor (which indeed accounts for overstrength), gives a direct indication of whether the design behaviour factor may be increased (analysis q larger than design q), decreased (analysis q less than design q) or unaltered (equal analysis and design q values).

Whereas recent work (Papazoglou and Elnashai, 1996, Elnashai and Papazoglou, 1997) has indicated that vertical earthquake motion may have a marked effect on seismic response, especially axial force-sensitive shear criteria, the criteria discussed above are assessed under horizontal earthquake ground motion only. The main reason for this is to enable comparison of results with studies undertaken by other research groups within the European collaborative research network Prenormative Research in Support of Eurocode 8 (PREC8). It is likely that the behaviour factors calculated for RC structures will be affected by the inclusion of vertical earthquake motion, but only if (*a*) an axial force-sensitive shear failure criterion is implemented and (*b*) natural records with ratio of vertical-to-horizontal more than two-thirds and absolute peak vertical ground motion more than 0·2–0·3 g are used.

5.3.4.3. Presentation of comparative results. Below, results from the analysis of the three configurations (twelve structures) are presented. The four EC8 spectrum-compatible artificially generated records discussed in section 5.3.3 above are referred to as EC8-1 to EC8-4. The results are presented in this section with general comments only; further discussion is given in section 5.3.4.4.

Observations from static push-over analyses are given in Table 5.3 below. It is noteworthy that the level of overstrength is substantial. This may be attributed in general to the use of mean values of material strength whereas the design utilizes characteristic strength levels. Also, increase in the amount of reinforcement and the use in design of partial material safety factors are contributors to the level of overstrength. Peculiar to this study is the use of 2-D analysis for 3-D designed structures. Therefore, the analysis does not include the effect of accidental eccentricity (torsional moments leading to additional column shear forces). Hence, the level of overstrength observed in Table 5.3 may be higher than a 3-D analysis would indicate.

Moreover, the top displacement corresponding to a 3% drift gives an indication of the uniformity of displacement distribution over the height. Hence, for the core structures, large displacements are observed, in common with the regular frame. The irregular frame seems to suffer from drift concentration, hence the top displacement is significantly lower in this case, compared to the other two configurations. The results shown in the table also indicate that the difference in top displacement is insignificant among structures designed to the same ground acceleration using different ductility classes.

Summarized in Table 5.4 below are the observed global and storey response parameters for all structures under the design and twice the design acceleration.

It is noticeable in Table 5.4 that the difference in overstrength among each design ground acceleration pair is larger for the design earthquake than for twice the design level. Furthermore, this difference is invariably higher between the DCM/DCL pair than for the DCH/DCM pair. With regard to deformations, the variations between structures designed to the same ground acceleration is rather small for the design event, increasing significantly for double the design acceleration. The expectation that higher ductility structures will have lower strength and will experience higher deformations is only partially confirmed; in several cases, lower ductility structures deform more than their high ductility counterparts.

Table 5.3. Overstrength from static analysis (Δ_m corresponds to 3% drift anywhere)

Reference	V/W (design)	V/W (observed)	Overstrength	Δ_y: mm	Δ_m: mm
Core-H030	0·201	0·569	2·83	170	652
Core-M030	0·261	0·691	2·65	196	648
Core-M015	0·130	0·355	2·73	156	656
Core-L015	0·193	0·465	2·41	187	664
Irr-H030	0·175	0·400	2·28	213	528
Irr-M030	0·233	0·504	2·16	232	528
Irr-M015	0·113	0·265	2·34	183	496
Irr-L015	0·169	0·327	1·93	199	500
Reg-H030	0·123	0·286	2·32	227	616
Reg-M030	0·160	0·352	2·20	276	648
Reg-M015	0·079	0·206	2·61	224	636
Reg-L015	0·119	0·280	2·35	249	644

Table 5.4. Analysis results for all structures at intensity 1·0 and 2·0 (average over four earthquake records)

	Base shear coefficient		Top displacement: mm	Inter-storey drift: %
	Average	Av./design	Average	Average over height
Intensity 1·0 (design)				
Core-H030	0·479	2·38	200·36	1·07
Core-M030	0·556	2·13	202·94	1·10
Core-M015	0·261	2·01	96·33	0·50
Core-L015	0·310	1·61	123·19	0·66
Irr-H030	0·336	1·92	233·27	1·30
Irr-M030	0·409	1·75	235·35	1·30
Irr-M015	0·214	1·89	138·01	0·78
Irr-L015	0·209	1·24	132·07	0·68
Reg-H030	0·265	2·15	247·27	0·85
Reg-M030	0·300	1·87	275·97	0·96
Reg-M015	0·148	1·87	169·67	0·55
Reg-L015	0·179	1·50	168·35	0·55
Intensity 2·0 (twice design)				
Core-H030	0·606	2·99	336·36	1·73
Core-M030	0·708	2·71	299·78	1·73
Core-M015	0·356	2·74	230·99	1·25
Core-L015	0·458	2·37	219·52	1·16
Irr-H030	0·387	2·21	435·11	2·51
Irr-M030	0·466	2·00	449·49	2·31
Irr-M015	0·252	2·23	231·78	1·39
Irr-L015	0·298	1·76	259·85	1·49
Reg-H030	0·301	2·45	490·84	1·75
Reg-M030	0·340	2·12	515·21	1·78
Reg-M015	0·218	2·76	311·63	1·10
Reg-L015	0·253	2·12	282·02	0·95

Average inter-storey drift values for the design and twice the design intensity for each set of the four structures are presented in Figs 5.35, 5.36 and 5.37, for Core, Irr and Reg, respectively.

The interrelationship between design ground acceleration and behaviour factor results in differing reinforcement detailing characteristics across the four structures of each configuration. In general, for two structures designed to the same ground acceleration, the structure associated with the larger behaviour factor is expected to possess reduced levels of longitudinal steel (lower seismic forces) in favour of more stringent ductility detailing

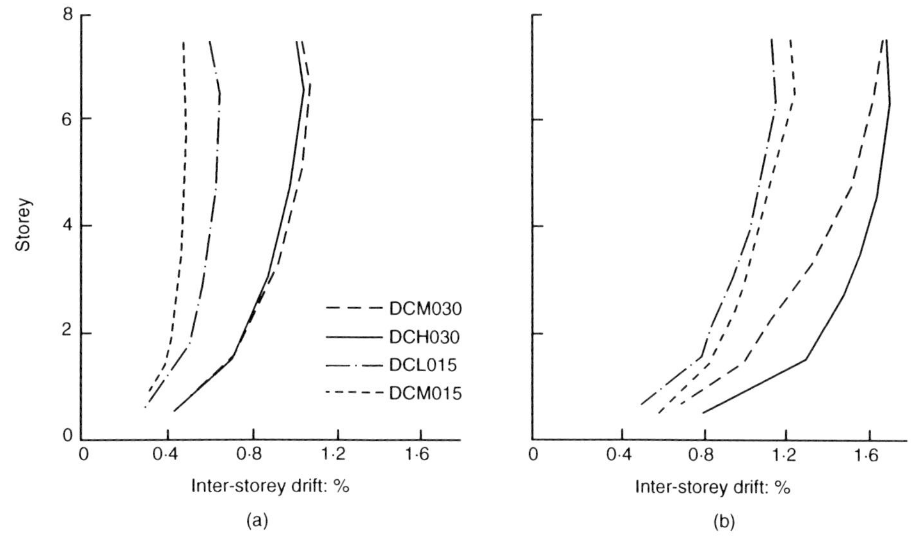

Fig. 5.35. Average drifts for all frame–wall structures at (a) the design and (b) twice the design ground acceleration

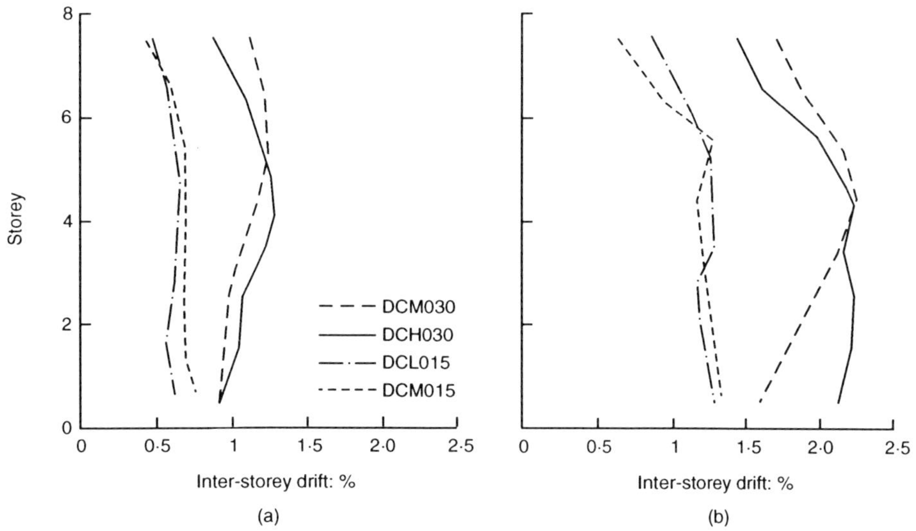

Fig. 5.36. Average drifts for all irregular frames at (a) the design and (b) twice the design ground acceleration

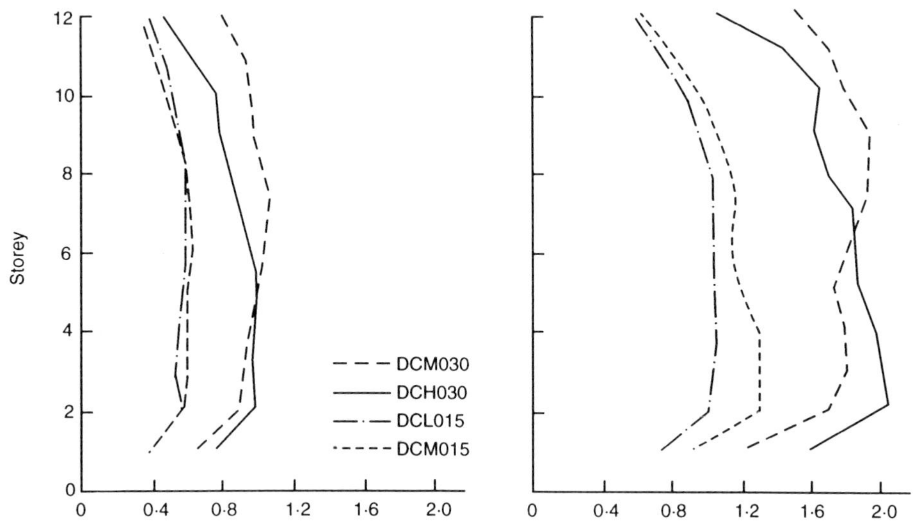

Fig. 5.37. Average drifts for all regular frames at (a) the design and (b) twice the design ground acceleration

Table 5.5. Curvature ductility demand (intensity 2·0 and EC8-1) — frame–wall

	Core-H030	Core-M030	Core-M015	Core-L015
External bay beams, average	3·55/2·24	3·01/2·19	2·93/1·60	1·84/1·10
Coupling beams, average	6·72	5·37	3·59	1·83
EC8 CCDF columns	13	9		5
EC8 CCDF walls	9·80	5·51		2·45
Column A, storey 1	1·43	1·48	—	—
Column B, storey 1	2·06	2·16	1·49	1·13
Column C, storey 1	1·31	1·23	—	—
Wall, storey 1	3·50	3·82	2·51	1·71

Table 5.6. Curvature ductility demand (intensity 2·0 and EC8-1) — irregular frames

	Irr-H030	Irr-M030	Irr-M015	Irr-L015
Ext. frame beam 1, storey 1	2·10/—	1·99/—	1·32/—	—/—
Ext. frame beam 2, storey 1	3·32/—	3·31/—	1·16/—	—/—
Ext. frame beam 3, storey 1	2·01/1·88	1·90/1·54	1·68/1·32	1·05/—
EC8 CCDF columns	13	9		5
Column A, storey 1	3·70	3·61	2·18	1·87
Column B, storey 1	3·73	3·85	2·09	1·79
Column C, storey 1	2·39	2·58	1·59	1·49
Column D, storey 1	2·99	2·92	1·81	1·81
Column E, storey 2	3·10	3·82	1·80	2·02

Table 5.7. Curvature ductility demand (intensity 2·0 and EC8-1) — regular frames

	Reg-H030	Reg-M030	Reg-M015	Reg-L015
Ext. frame beam 1, storey 1	2·74/—	2·33/—	1·72/—	—/—
Ext. frame beam 2, storey 1	3·52/—	3·41/—	1·21/—	—/—
Ext. frame beam 3, storey 1	2·18/2·05	1·96/1·64	1·72/1·40	1·05/—
EC8 CCDF columns	13	9		5
Column A, storey 1	4·03	2·59	2·19	1·24
Column B, storey 1	4·03	2·64	2·16	1·24
Column C, storey 1	2·10	1·63	1·27	—
Column D, storey 1	2·27	1·76	1·38	—
Column E, storey 1	3·00	2·34	1·64	1·10

requirements (higher levels of transverse reinforcement). Tables 5.5, 5.6 and 5.7 present curvature ductility demand for the selected members shown in Figs 5.24, 5.25 and 5.26 of section 5.3.2. In order to account for the variation in axial load in vertical members, the yield curvature is evaluated at the level of axial load corresponding to the maximum curvature. Curvature ductility for positive and negative beam moments are reported.

It is observed that member curvature ductility demand is on the whole rather modest, compared to the EC8 target values. The irregularity of having a planted column in the 8-storey frame caused only marginal effect on the more constrained of the beams supporting the column, but global response was largely unaffected.

Figure 5.38 presents values of behaviour factors obtained by progressively scaling the four input motions until the attainment of pre-defined yield and collapse limit states (Salvitti and Elnashai, 1995). The minimum q' values presented correspond exclusively to (*a*) the exceedance of a yield top displacement (defined by assuming an equivalent elasto-plastic system with reduced stiffness, evaluated as a secant passing through 75% of the maximum load) and (*b*) the attainment of 3% drift at any storey level.

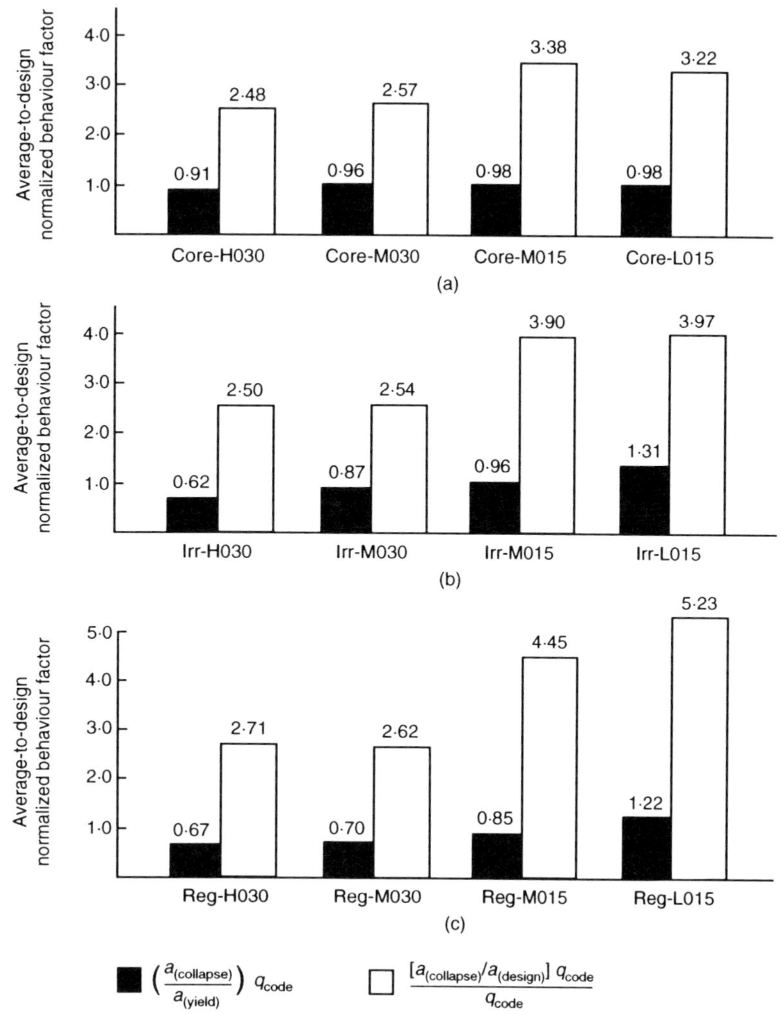

Fig. 5.38. Minimum behaviour factors (normalized by design behaviour factor) for (a) frame–wall; (b) irregular frames; (c) regular frames

The normalised q'_D values reported in Fig. 5.38 (light shade), obtained by the quotient of the collapse and code-implied yield accelerations (given by the design acceleration divided by q), implicitly account for overstrength at the yield level, as discussed above. As such, it is a more meaningful measure for the assessment of the behaviour factor used in design. It is clear from Fig. 5.38 that there is room for increasing the design behaviour factor in EC8. This is discussed further below.

5.3.4.4. Discussion of the results. The large amount of performance criteria monitored during analysis sheds light on the effect of various capacity design and detailing characteristics on the inelastic response of the structures. For a rigorous assessment of the results presented above, these should be viewed in tandem with the differences between various EC8 ductility class regulations given in the Appendix to this Chapter. Hereafter, a sample of the results is discussed with a view to highlighting the comparative seismic performance of structures designed to the same ground acceleration using different sets of capacity design recommendations.

5.3.4.4.1. Frame–wall structures (8 storeys). At the design level, comparison between global response of the DCM030 and DCH030 structures gives a clear picture, whereby the former has a lower strength level corresponding to practically the same displacement, as reported in Table 5.4 (2nd column). This, when viewed alongside Fig. 5.35(a), indicates

Table 5.8. Comparisons of forces and deformations

Intensity = 1·0	Av./design base shear	Top displacement	Drift
H030/M030	1·11	0·98	0·96
M015/L015	1·24	0·78	0·75
Intensity = 2·0			
H030/M030	1·10	1·12	1·00
M015/L015	1·15	1·05	1·07

that there are indeed differences in global response, but these are rather small. Furthermore, the trend in base shear is contrary to expectations, which may be attributed to the dominant effect of the core. This point is emphasized in the same figure, where significantly larger differences in drift are observed for the 0·15 g design acceleration pair of structures. Interestingly, the lower ductility class structure (DCL015) experiences higher levels of deformation, as reported in Table 5.4.

The same observation applies, to a lesser extent, to the global performance under twice the design acceleration, also summarized in Table 5.4, where average inter-storey drifts are presented for the 0·30 g and 0·15 g structures, respectively, and Fig. 5.35(b). Although, on average, the higher ductility class structures (DCH030 and DCM015) experience greater levels of deformation, the differences with respect to their design acceleration partners are small. The comparison between the four structures in terms of drift, top displacement and average-to-design base shear is summarized in Table 5.8.

Table 5.8 demonstrates that for twice the design intensity, the difference between base shear ratios for the DCH030 and DCM030 structures is 10%, while the corresponding value for the 0·15 g structures is 15%. The same is observed in the overstrength from static analysis presented in Table 5.3 where the difference between the DCH030–DCM030 structures is 6% while that for the DCM015–DCL015 structures is 13%. An interesting parallel comparison can also be made with reference to the minimum average-to-design behaviour factors (q'_D) reported in Fig. 5.38(a). It is noted that the obtained q'_D values follow the same trend as the average-to-design base shears presented in Table 5.8. The implication of this is that the stringent application of ductility class 'H' rules in this type of structure is not reflected in the seismic performance.

In comparing local demand in beams shown in Table 5.5, smaller differences are observed between the 0·30 g structures. Such behaviour is justified by comparing the design q-factor ratio of the two design acceleration pairs (1·50 for the 0·15 g structures as compared to 1·33 for the 0·30 g structures). It is observed that the increase in bi-diagonal reinforcement for the DCM030 structure of approximately 30% leads to a reduction in demand of 25%, on average. For walls, an increase in longitudinal reinforcement of 50% and a reduction in hoop steel of 30% result in an increase in demand of 9%. A higher maximum curvature for the DCM030 structure is observed at a higher level of axial load with a correspondingly higher yield curvature. The relative increase in maximum curvature between the DCH030 structure and the DCM030 is higher than the relative increase in yield curvature resulting in the observed reduction in demand. In both structures, inelasticity extended up to the fifth level.

For the DCM015 and DCL015 comparison, it is noted that beams and columns behave in a similar manner. For the walls and coupling beams, it is clear that the increase in bi-diagonal steel and wall longitudinal reinforcement has an unfavourable effect on the local inelastic response which was observed in the first and second storeys in both structures. The influence of gravity loading in column design, illustrated by small differences in longitudinal reinforcement in Column A (corner) and Column C (joining long span beams), results in elastic response under twice the design earthquake for both structures.

All results presented indicate that the ductility demand for beams and columns is rather low, while that for coupling beams and walls is moderate. Indeed, the response of the structure is dictated by the wall behaviour, as confirmed by the inter-storey drift plots, while the input energy is mainly dissipated in the coupling beams.

5.3.4.4.2. Irregular frames (8 storeys). Table 5.6 and Fig. 5.36(a) report the average global response parameters at the design intensity. It is observed that for the 0·30 g design acceleration cases, lower base shear forces are associated with the DCH030 structure with less marked difference in deformation both at the global and storey levels. Although the base shear forces attracted by the DCM030 structure are higher in numerical terms, the average-to-design ratio is lower. The same observation is made for the 0·15 g design acceleration comparison with respect to the forces and deformations of the DCL structure.

In comparing the global response of the 0·30 g design acceleration cases, presented in Fig. 5.36(b) for twice the design intensity, higher levels of drift are observed for the DCH structure. Although the global displacement response almost coincides with that of the DCM design for the intermediate storeys, as shown in Fig. 5.36(b), higher drift values for the former structure in the lower storeys are caused by the interaction of local irregularities and lower design forces.

Among structures designed to the same ground acceleration, that which is associated with a larger behaviour factor is expected to possess greater ductility and hence to undergo higher levels of deformation. In considering the 0·15 g designs at twice the design acceleration it is observed that the DCM structure attracts lower forces, as expected. However, it experiences lower levels of deformation both at the global and storey levels, as shown in Fig. 5.36(b). This observation demonstrates that inelastic response cannot be represented solely by the behaviour factor, as a consequence of the interaction between various design parameters. Table 5.9 summarizes the observed global response parameters for the four irregular frames considered.

Comparison of average-to-design base shears presented in Table 5.9 with the obtained minimum q'_D values yields the same conclusion reached for the frame–wall structures. Further, the ratios given in Table 5.9 for the 0·30 g pair and the 0·15 g pair confirm that the difference in response characteristics between DCM and DCL is significantly higher than for the DCH and DCM. The effect of irregularity on local response of all structures can be observed in Table 5.6. Of particular interest is the distribution of demand at the external bay intermediate joint. The absence of ground-storey

Table 5.9. Comparison of forces and deformations

Intensity = 1·0	Av./design base shear	Top displacement	Drift
H030/M030	1·10	0·99	1·00
M015/L015	1·52	1·04	1·15
Intensity = 2·0			
H030/M030	1·11	0·97	1·09
M015/L015	1·27	0·89	0·93

columns at such locations resulted in a redistribution of demand to neighbouring beams and columns.

Although external frame beams 1 and 2 (joining the planted column) are of equal span to beam 3, demand observed in the former members exceeds that in the latter (Table 5.6) in all cases except for the DCL015 structure, where behaviour remains near-elastic. Demand in beams in the intermediate levels is comparable, at least for the 0·30 g design, to those obtained in the higher storey beams in the frame–wall structures. This observation confirms that horizontal elements provide the primary source of energy dissipation in such structures. Further, the beneficial effects of reducing longitudinal steel in favour of higher levels of transverse reinforcement is achieved, in relative terms, by selecting DCM for a low design acceleration.

Hinging of the 'planted' column (Column E) base occurred in all cases. Although below the code-prescribed ductility supply values (CCDF in Table 5.6), demand for all columns is higher than that observed for frame–wall members highlighting the favourable effect of the coupled walls. In comparing column demand for the 0·15 g design structures, the obtained values are consistent with ductility class philosophy, i.e. higher values are observed for the higher DC structure considered, with the exception of the planted column. The same observation applies in comparing demand for the 0·30 g designs.

Within each design acceleration pair, the planted columns in the lower DC structures (DCM030 and DCL015) attain lower maximum curvatures at lower levels of axial load for the input motion considered. Hence, although the corresponding yield curvatures are lower, the resulting demand is higher.

5.3.4.4.3. Regular frame (12 storeys). Analysis results for the design intensity are presented in Table 5.4 and Fig. 5.37(a). As for the irregular frames, lower base shear forces are attracted by the higher DC structures (H030 and M015). Although these are higher in numerical terms, the average-to-design values are lower. As for the frame–wall and irregular frame configurations, it is observed that, for the higher acceleration pair, greater levels of deformation are experienced by the lower DC structure. Comparison of global displacement response for the 0·30 g design acceleration pair, shown in Fig. 5.37, illustrates a fundamental difference in the distribution of inter-storey drift. Whereas the DCH structures experience greater deformations in the lower storeys, higher drifts are observed in the intermediate–high storeys of the M030 structure. From the same figure it is observed that deformation response of the lower acceleration pair is virtually identical, both at the global and storey levels.

In comparing the two frame configurations at the design intensity, it is observed that, although top displacements are higher for the regular frames, the irregular structures experience greater levels of inter-storey drift, irrespective of design acceleration and DC.

Response parameters for twice the design intensity are presented in Table 5.4 and Fig. 5.37(b). In comparing the performance of the 0·30 g design acceleration pair, the same observation is made as for the irregular configuration. The M030 structure undergoes greater deformations both at the global and storey levels. Designing high-rise frames to a higher DC in low seismicity areas shows its benefits in Table 5.4.

From the comparison shown in Table 5.10, which gives ratios of response parameters for pairs of structures, it is confirmed that the difference in behaviour is more marked for the lower acceleration pair, particularly at high intensities.

Table 5.10. Comparisons of forces and deformations

Intensity = 1·0	Av./design base shear	Top displacement	Drift
H030/M030	1·15	0·90	0·83
M015/L015	1·25	1·01	1·00
Intensity = 2·0			
H030/M030	1·15	0·95	0·98
M015/L015	1·30	1·10	1·16

As expected, beam demand, shown in Table 5.7, follows the pattern intended by the code, that is more frequent excursions into the inelastic range for the higher DC structures (H030 and M015). A direct comparison with irregular frame beams would only be possible with a set of frames of equal height. In general, the most significant differences are observed for the 0·30 g design pair. It is noteworthy that no comparison can be made between the obtained demand code suggested values since these are not dealt with in EC8.

Table 5.7 also presents values of curvature ductility demand for the selected vertical members. A comparison with irregular frame elements furnishes two observations. The first is the close agreement between demand in Columns A to D of Table 5.6. The interrelationship between levels of longitudinal reinforcement and axial load variation with observed demand is confirmed. The second observation is made with regard to the effect of irregularity on demand. By comparing demand imposed on Column E in Tables 5.6 and 5.7, it is apparent that these are higher in the cut-off element, particularly for the lower acceleration pair.

5.3.4.5. Concluding remarks. Many of the differences between ductility classes summarised in the Appendix to this chapter apply concurrently in the cases studied. It is therefore difficult to ascribe specific behavioural patterns to specific capacity design characteristics. Notwithstanding the complexity of inelastic response of the buildings considered, in view of the ductility-based design and detailing criteria, the following conclusions are supported by the results given.

(*a*) In all cases, capacity design criteria successfully account for plastic hinge formation at beam ends. The following hierarchy of element hinging, in the frame–wall structures considered, is typically observed

- coupling beams in the top two storeys
- coupling beams in the bottom one or two storeys
- lower storey walls and
- top-storey beams
- ground-storey columns
- mid-height beams and coupling beams
- lower storey beams.

(*b*) Hinging at ground-storey column bases was realized by application of capacity design detailing rules for all configurations considered, but no mechanisms were observed.

(*c*) Clear correlation exists between ductility class detailing rules and the observed behaviour of both beams and coupling beams, particularly for the latter elements which provide the primary source of energy dissipation in frame–wall structures. Their more satisfactory performance in the higher ductility class structures is partnered with lower levels of bi-diagonal longitudinal reinforcement and therefore greater ease of construction. In beams, the rather low demand observed is an indication that seismic loading is not the dominant design scenario.

(*d*) Capacity design rules in columns show a less clear correlation with the observed inelastic response. In the higher ductility class structures (DCH030 and DCM015), the higher hoop volumetric ratios in elements furthest from the walls do not lead to noticeable differences in behaviour. However, significantly higher demand is associated with the lower levels of longitudinal reinforcement in columns closest to the walls. Ductility class requirements appear to have practically no effect on the performance of vertical elements in the lower design acceleration structures, the design of which is gravity loading dominated.

(*e*) Ductility demand imposed on columns in frame–wall structures is rather modest. It is therefore reasonable to release capacity design requirements for column overstrength in structures whose behaviour is dominated by walls. Furthermore, with reference to the distribution of demand in beams (increasing with height), this release is more applicable to columns at lower storeys than elsewhere.

(*f*) It is confirmed that frame–wall systems are successful in satisfying drift requirements, regardless of ductility class, design acceleration and intensity. As expected, higher levels of drift were observed for moment-resisting frames designed to high accelerations. The combined effect of geometric irregularity and design behaviour factor (ductility class) on deformation response was most critical in lower storeys for DCH, although only for twice the design intensity.

(*g*) It is apparent that irregularity at a joint induces higher demand on neighbouring beams and columns. Increase in demand corresponds to lower levels of longitudinal reinforcement in favour of closer hoop spacing in critical regions of columns and beams. In comparing demand for the two structural configurations, it is observed that coupled walls serve to protect columns, emphasizing the point made in (*e*) above.

(*h*) In walls, lower levels of longitudinal steel and higher hoop volumetric ratios in boundary elements do not necessarily lead to higher demand, as a result of the high variation in axial load. Although higher axial loads lead to increases in yield curvature, higher maximum curvatures are observed as a result of the longitudinal steel distribution within the section. It is also observed that within each design acceleration pair inelasticity spreads equally along the height of the wall. The design implication of this observation is that for high design accelerations, detailing of critical zones in walls should extend to more than half the building height, even for DCM.

(*i*) Overall, the results for the three structural configurations considered show that, although changes in capacity design requirements do lead to local differences in behaviour, significantly fewer marked effects are observed in global response. It is also clear that for a given design acceleration, design to the lower ductility class is more cost-effective. Thus, similar levels of performance are achieved, avoiding more stringent detailing requirements. This conclusion applies more to DCH and DCM than to DCM and DCL.

On the whole, the analysis of the twelve structures described above (which belong to category (*b*) described in section 5.3.1) furnished an understanding of the effect of the application of a set of capacity design criteria and their interrelationship with the static load design level. Further studies according to the category (*a*) approach (variation of a single capacity design parameter) would lead to quantitative measures of parameter–response pairs. Examples of such studies are presented in section 5.2.

5.3.5. *Inelastic response of multi-storey buildings designed according to different column capacity design approaches, as affected by the width of slabs effective in tension*

5.3.5.1. Scope of inelastic response analyses. In the second part of this case study the inelastic seismic response of 4-storey, 12-storey and 3-storey frame structures is computed for the four artificial motions applied separately in the two horizontal directions X and Y. The motions are scaled to 1·0, 1·5 and 2·0 times the design intensity of each structure.

The inelastic response is computed both for the structures with columns designed according to the 'relaxed' capacity design procedure of Eurocode 8 and of equation (2.14) of this Design Guide, as well as for those with columns proportioned according to the more demanding full capacity design approach of equation (2.12). As no capacity design magnification of column moments is required for the DCL structures, and as in the DCM 12-storey structures designed for a 0·15 g ground acceleration the two capacity design approaches give identical column reinforcement, the second capacity design alternative needs to be considered only for eight among the total of twelve 4-storey, 12-storey or 3-storey structures, giving a total of 20 buildings to be analysed in both horizontal directions, for four ground motions applied at three intensities each.

As the results of Fig. 5.28 show, the relaxed capacity design procedure of equation (2.14) provides *per se* little column overstrength relative to the design seismic action effects computed from the linear elastic analysis of the structure. Significant column overstrength is derived, however, from the minimum reinforcement requirements, in combination with the relatively large column dimensions selected for drift control and anchorage of beam bars in the joints. On the other hand, although a very small part of the slab reinforcement has been considered to contribute to the beam top reinforcement during proportioning, it is now widely recognized that a very significant amount of the slab reinforcement which is parallel to a beam fully contributes to its flexural capacity, especially when the beam response is well beyond yielding. As a result, beam yield moments in negative bending (top flange in tension) significantly exceed the beam flexural capacities M_{Rb} considered in the capacity design calculations according to equations (2.12) or (2.14) of section 2.4.3.2. For this reason, and despite the aforementioned column overstrengths, column plastic hinging during the actual inelastic response is not unlikely. This is more likely for the designs which follow the relaxed capacity design format of equation (2.14). In other words, the fact that beams are in reality stronger than considered in the capacity design based proportioning of columns, works against our efforts to control inelastic response by forcing plastic hinging to take place in the beams rather than in the columns. To investigate this question, a second set of inelastic seismic response analyses is performed, in which a large part of the slab reinforcement is considered to participate fully in beam bending. In these alternative calculations, the effective in-tension width of the slab is (conservatively for this purpose) taken to extend on each side from the beam axis up to a distance equal to half the span of the beam or the span of the slab normal to the beam axis, whichever is smaller (i.e. equal to the smaller dimension of the slab panel on the side of the beam axis where the effective slab width is calculated). This value is close to an upper limit for the effective slab width, an upper limit which is very likely to be reached under strong ground motions that may drive the beam in negative bending well into the inelastic range.

The set of analyses representing the structures as designed and as intended with regard to the beam–column strength ratio, is identified with the term 'zero slab width', while that of the structures as they are most likely to be in reality, is identified with the term 'full slab width'. The total number of seismic response analyses is 960.

5.3.5.2. Modelling approach and assumptions. A simple member-type lumped inelasticity model is adopted for the beams and the columns of the frame structures. The model employs a (end) moment–chord rotation relation at each member end which is based on the assumption of antisymmetric bending and neglects the variation of (column) axial force during the response. In monotonic loading this moment–chord rotation (M–θ) relation is bilinear, with pre-yielding and post-yielding stiffnesses calculated on the basis of the Park and Ang (1985) model for the chord rotation at yield θ_y and for the ultimate flexural capacity M_u of the member, and of the Park *et al.* (1987) model for the ultimate chord rotation θ_u. These parameters, and hence the M–θ model, include, albeit in gross approximation, the effect of shear deformations within the shear span and the contribution of bar slippage within the joint beyond the member end. For cyclic loading, the nine hysteresis rules of the Takeda (1971) model, as simplified by Otani (1974) and Litton (1975), are used. It is also noteworthy that although the structures are analysed in 3-D, with simultaneous column bending considered in both horizontal directions, the response is effectively one-directional, as the ground motion is applied separately in the two horizontal directions and the structures are symmetric with respect to both horizontal axes.

Floor slabs are considered as rigid diaphragms, and second-order (P–δ) effects are included in the model.

Damage is expressed through an energy-based damage index proposed by Fardis (1994), which assumes the value 1 or 100% at loss of the capacity of the member to support gravity loads, i.e. (well) beyond ultimate strength. The main failure parameter employed by this damage index is the ultimate chord–rotation θ_u computed according to Park *et al.* (1987).

Despite the simplicity of the modelling, global and local response predictions obtained therefrom have been validated (Fardis and Panagiotakos, 1997) through comparisons with results of quasi-static or pseudodynamic tests on full-scale structures or subassemblies. Moreover, computed average storey drifts for the four artificial motions applied in direction X of the 12-storey structures at intensities 1·0 and 2·0 are up to 20% higher than those computed in section 5.3.4 using a much more sophisticated and detailed model. The difference between the two modelling approaches is also justified, since the simple models used in this section include, albeit in gross approximation, bar slip effects within the joints and member shear deformations, which are neglected in the model of section 5.3.4.

The mean material strengths are considered for the members of the as-built structures: a yield stress of 585 MPa for the S500 steel and a compressive strength of 33 MPa for the C25/30 concrete. These mean strengths are about 4/3 and twice the design strengths of steel and concrete, respectively.

5.3.5.3. Storey sway potential of the as-intended and the as-built structures. Priestley and Calvi (1991) proposed the sway-potential index S_p as a measure of the tendency for soft-storey development in one horizontal direction. They define this index as the ratio of the sum of beam flexural strengths to that of column flexural strengths in the storey and the horizontal direction of interest, both sums extending over all joints of the storey (positive beam strength in one side of the joint plus negative on the other, and corresponding column strengths). The value of the S_p index per storey and horizontal direction is computed herein on the basis of the member yield strengths for all four combinations of capacity design approach — equation (2.12) versus equation (2.14) — with effective slab width — 'zero' versus 'full' — and presented in Fig. 5.40 for the 'conventional' capacity design of equation (2.12) and Fig. 5.39 for the 'relaxed' of equation (2.14). These results, to be used later on in the evaluation of the non-linear response computed for these four combinations, show the following.

(*a*) As expected, the value of S_p (i.e. the tendency for soft-storey development) decreases, as the ductility class increases. However, the difference is more systematic and larger between DCL and DCM than between DCM and DCH, and increases when the conventional capacity design, equation (2.12), is used instead of the relaxed, equation (2.14). The small and non-systematic difference between DCH and DCM is noteworthy.

(*b*) With the exception of the top storey of the low-rise buildings and of the DCL 3-storey structure in the long-span direction Y, in which gravity loads control the design of the beams, the values of S_p are consistently below 1·0, showing almost no tendency for soft-storey development, irrespective of the capacity design approach used and of the effect of the slab width in tension.

(*c*) When the effective slab width in tension is considered, the value of S_p increases, but not dramatically so (depending, of course, on the slab–beam tension steel ratio).

(*d*) The main effect of relaxing capacity design according to equation (2.14) is to reduce the difference between the S_p values of the different DCs, especially those of DCM and DCH.

The reader is cautioned that these results refer to the yield strengths of the members, used for signalling plastic hinge formation in the non-linear response analyses, and not to the flexural ultimate capacities, the design values of which are employed in the application of capacity design in the design phase. If S_p is calculated, instead, on the basis of the design ultimate strengths of beam and columns, its value decreases by as much as 17% in most cases, but increases sometimes by up to 10% relative to its value based on mean yield strengths. The difference between the mean and the design value of the material strength is much larger for concrete (~ 2) than for steel ($\sim 4/3$) and causes a larger difference between mean yield and design ultimate strength in the columns than in the beams, explaining an increase in the value of S_p when design ultimate strengths are used instead of mean yield values. However, for given material strengths the difference between the yield strength (identified with first yielding of a tension bar) and the ultimate is lower in the beams, especially in positive bending (slab in compression), due to the concentration of the steel near the extreme fibres. This difference causes a reduction in the value of S_p when going from yield strengths to ultimate, a reduction which seems to dominate over the increase due to the material strength values. Accordingly, S_p values computed in the design phase on the basis of the member design flexural capacities would

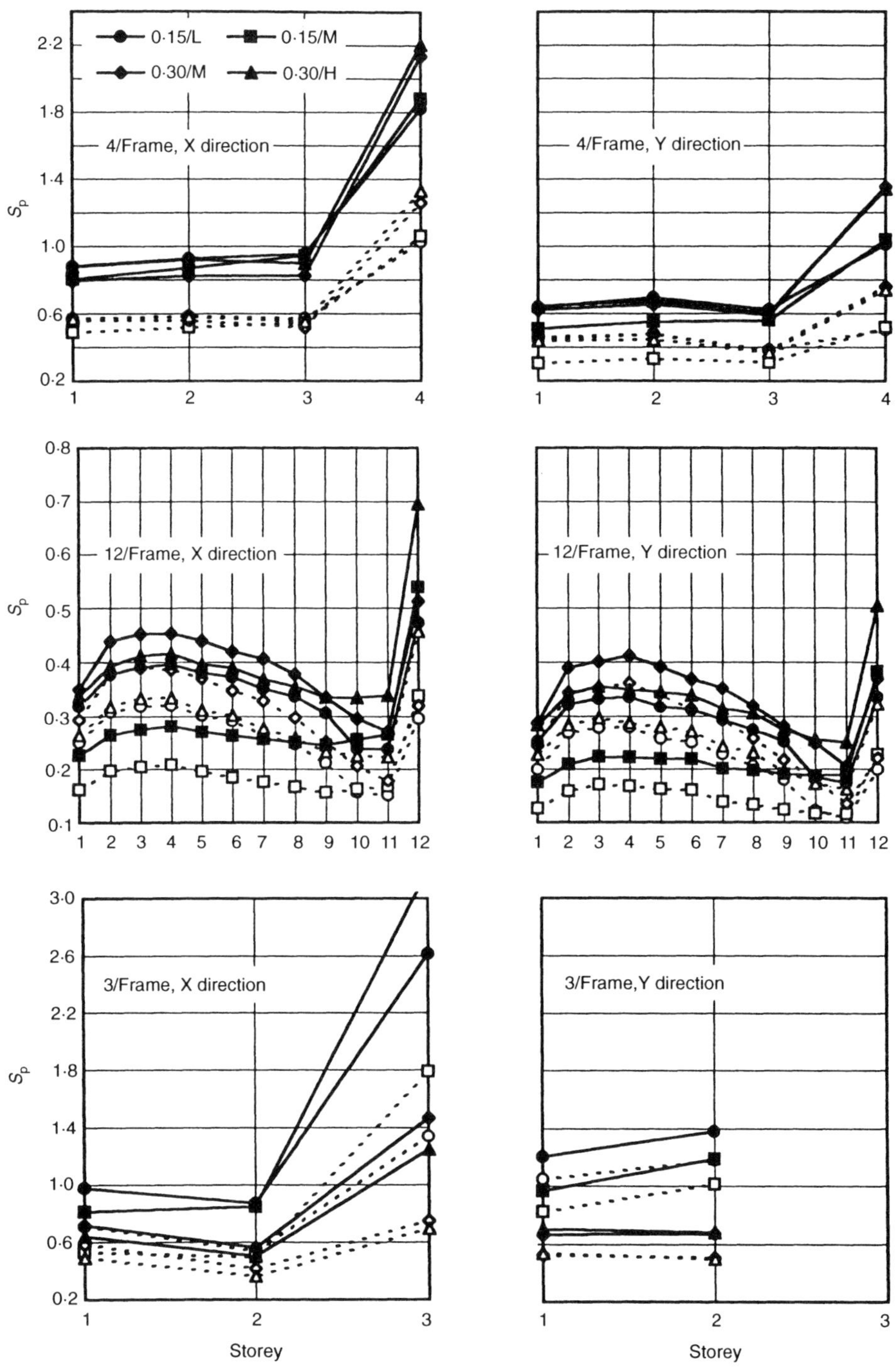

Fig. 5.39. Sway potential index of structures designed according to equation (2.14), with relaxed capacity design. Full symbols and solid lines: 'full slab'; empty symbols and dotted lines: 'zero slab'

have been overall slightly lower than those shown by the dashed line in Figs 5.39 and 5.40, suggesting even lower tendency for soft-storey development.

5.3.5.4. Presentation and discussion of inelastic response results. Tables 5.11 to 5.16 present average, over the four artificial motions, global response results for the 4-storey, 12-storey and 3-storey structures. Results are only presented for the 'full slab width' case of the structures as they are in reality. Those for the 'zero slab width' case, i.e. for the structures as conceived in design, are only commented upon to show which the real or the as-intended structures meet the design aim for controlled inelastic response best. The tables are presented in pairs, one for each type of structure, with the first pair referring to the structures designed according to

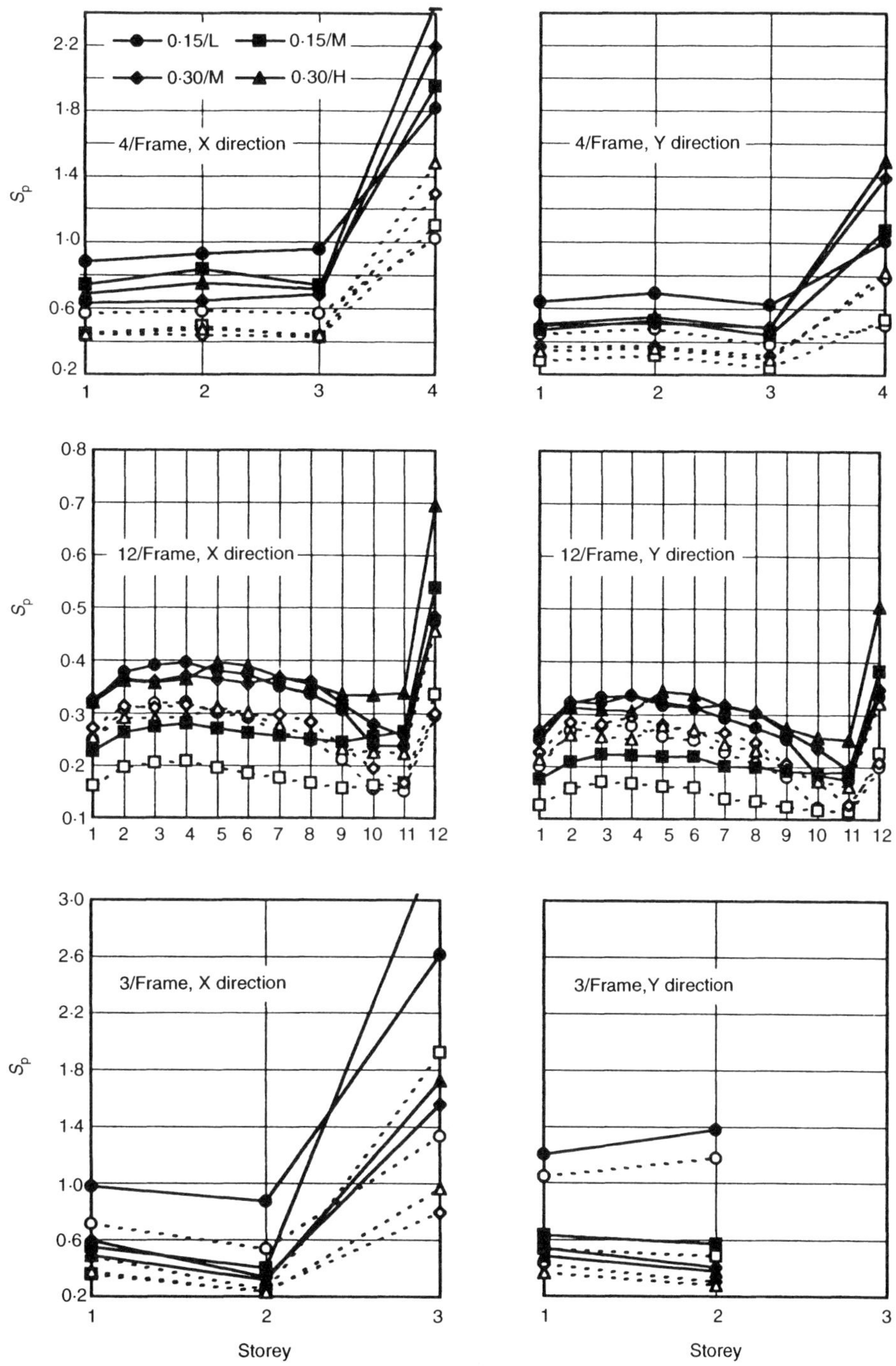

Fig. 5.40. Sway potential index of structures designed according to equation (2.12), with conventional capacity design. Full symbols and solid lines: 'full slab'; empty symbols and dotted lines: 'zero slab'

the relaxed capacity design procedure of equation (2.14) and the second to the designs following the full capacity design of equation (2.12).

All conclusions mentioned in the following regarding differences between the structures in which the full capacity design, equation (2.12), is applied, and those following the relaxed capacity design format of equation (2.14), apply throughout all three groups of structures. However, these differences are least significant in the group of 12-storey structures. The reason is that in this class of building the effect of higher moment magnification in the columns of the full capacity design structures is overshadowed by the column overstrength due to the large column dimensions and the associated minimum reinforcement. These overstrengths are such that in the DCM 12-storey structures designed for a 0·15 g ground acceleration, the two capacity design procedures lead to the same column reinforcement.

The predominant period of the non-linear response T_{nl} listed in the fifth column of the tables is calculated from a Fourier analysis of the top displacement time-history. For the structures designed according to equation (2.14), T_{nl} is, on average, 10% longer than the fundamental elastic period T_{cr} of the cracked structure, with the stiffness of all members taken as equal to the secant stiffness at yielding. In turn this 'cracked' period is, on average, about 2·5 times the period of the uncracked structure, on which the design was based. At an excitation intensity of 2·0, the predominant period of the non-linear response of the structures designed according to equation (2.14), is about 25% longer than that of the cracked structure. The lengthening of the period, firstly and mainly due to cracking and then due to the inelastic action of the response, considerably reduces the seismic force demands on the structure, relative to those for which it has been designed: the lengthening of the fundamental period due to full cracking corresponds to a reduction of seismic action effects by about 45% on average for the structures designed for a 0·15 g acceleration, or about 40% for those designed for 0·30 g. The further lengthening of the predominant period due to the inelasticity that develops at an excitation intensity of 1·0 corresponds to a total average reduction of the ordinates of the 5% damped spectrum to about 55% of those for which the structures were designed, or to about 50% of the latter at an excitation intensity of 2·0. Therefore, for the twelve frame structures in the two horizontal directions this shift of the period (mainly) due to cracking, is roughly equivalent to a reduction in the q-factor for which the structures have been designed by a factor of about 1·8 for ground motions at the design intensity, or of about 2·0 at twice the design intensity. This 'understress' or 'overstrength on the demand side' factor of 1·8 to 2·0, includes only the effects of softening and not those of the increased energy dissipation during inelastic response. For the structures designed with full capacity design, equation (2.12) of section 2.4.3.2, the effective non-linear period of vibration T_{nl} is shorter, and the 'understress' due to the softening of the structure upon yielding is less pronounced.

When the contribution of slab reinforcement to the beam strength and stiffness is neglected, the predominant period of the non-linear response T_{nl} and that of the elastic but fully cracked structure T_{cr} increase, by about 10% on average, in comparison with the realistic case of the full slab width participation. However, this lengthening of the period is not real, and hence it is of interest only to the extent that it explains other features of the computed response, such as the increase in lateral drifts, etc. An interesting by-product of the shift of the predominant period further down the falling branch of the elastic spectrum of the input motion, is that the second and third translational modes, which now have effective periods in the constant acceleration plateau of the spectrum, may gain some importance in comparison with the first mode.

The penalty that the structure pays for the softening and the associated reduction in force demands is a very large increase in peak lateral drifts, both at the top and at the inter-storey level (10th and 11th columns of Tables 5.11–5.16). Drift ratios at the design intensity are from 2·5 to 5 times the elastic values calculated during design on the basis of the stiffnesses of gross uncracked sections. A large fraction of the inter-storey drift ratios, up to one-third of the total, is estimated to be due to slippage of bars in the joints. The rest is equal to the sum of the average beam plus column chord rotations in the storey. In the more realistic 'full slab width' case, in the low-rise buildings the maximum inter-storey drift takes place where it was

Table 5.11. Inelastic response results of 4-storey structures (average for four motions) — 'full slab width' — relaxed capacity design equation (2.14)

DC	a_g	Exc.	Period: sec		V_b/W: %		$\sqrt{(E_{in}/M)}$:	E_h/E_{in}:	Drift ratio: %			$E_{h,col}/E_{in}$: %		$E_{h,beams}/E_{in}$: %				Damage index: %, storey mean/max.							
																		Columns				Beams			
			T_{cr}	T_{nl}	Ult.	NL	m/s	%	Top	Inter.	Int./Top	1st	Rest/avg.	1st	2nd	3rd	4th	1st		2nd		1st		2nd	
Direction X																									
L	0·15 g	×1·0	1·21	1·24	32	14·4	0·39	45	0·72	0·89	1·23	7·6	6·5	5·4	4·6	3·5	4·6	1	3	1	4	1	4	1	4
		×1·5		1·25		20·8	0·58	42	1·09	1·34	1·24	6·6	5·3	6·9	5·8	4·5	1·9	3	6	3	7	3	6	3	6
		×2·0		1·30		26·7	0·76	42	1·45	1·83	1·26	6·3	4·6	8·4	6·9	5·2	1·2	4	11	4	10	4	8	5	9
M	0·15 g	×1·0	1·20	1·24	32	14·2	0·39	47	0·71	0·91	1·28	7·1	6·6	6·7	5·3	3·5	4·5	1	2	1	3	1	3	1	3
		×1·5		1·26		20·5	0·57	44	1·08	1·40	1·30	6·2	5·3	8·7	6·9	4·4	1·9	2	4	2	5	2	4	2	5
		×2·0		1·32		26·4	0·75	44	1·42	1·86	1·31	6·5	4·8	9·3	7·8	5·1	1·2	3	7	3	7	4	6	4	7
M	0·30 g	×1·0	1·08	1·14	41	29·8	0·79	42	1·21	1·52	1·26	6·3	4·9	7·1	7·0	5·4	1·1	2	5	2	5	3	5	3	6
		×1·5		1·28		38·8	1·12	50	1·94	2·56	1·32	8·2	4·8	10·3	10·1	6·1	0·8	4	11	4	10	5	8	5	10
		×2·0		1·50		43·5	1·45	54	2·50	3·36	1·34	11·5	5·5	10·3	9·6	5·6	0·7	6	17	6	17	8	14	7	15
H	0·30 g	×1·0	1·13	1·23	32	29·2	0·77	42	1·27	1·72	1·35	6·7	4·9	8·9	7·3	4·0	1·0	2	6	3	6	3	6	3	6
		×1·5		1·30		36·1	1·10	52	2·03	3·00	1·48	9·1	6·0	10·4	9·3	4·3	0·6	4	9	4	9	5	10	4	9
		×2·0		1·62		38·9	1·42	58	2·43	3·93	1·61	12·9	7·5	9·8	8·0	4·0	0·5	5	12	6	14	7	17	5	12
Direction Y																									
L	0·15 g	×1·0	1·33	1·37	28	13·4	0·38	48	0·79	0·98	1·23	6·2	4·5	7·4	6·0	8·3	6·5	1	2	1	2	1	2	1	2
		×1·5		1·41		19·0	0·57	45	1·11	1·40	1·27	6·2	4·1	8·3	6·7	7·5	4·2	2	5	2	4	2	4	2	4
		×2·0		1·55		24·5	0·75	45	1·39	1·86	1·34	6·7	3·7	9·3	7·5	6·9	3·5	3	9	2	5	5	8	4	7
M	0·15 g	×1·0	1·34	1·42	25	12·0	0·38	56	0·69	0·96	1·39	6·9	4·4	11·1	9·9	8·5	6·5	1	2	1	1	1	2	1	2
		×1·5		1·55		17·0	0·56	51	1·03	1·47	1·42	7·7	4·1	10·5	9·5	7·3	4·0	1	3	1	2	2	4	2	4
		×2·0		1·59		21·4	0·73	48	1·39	2·00	1·43	7·8	3·7	10·0	9·1	7·1	3·4	2	6	2	4	4	7	4	6
M	0·30 g	×1·0	1·20	1·32	30	28·2	0·76	44	1·36	1·74	1·28	6·5	3·8	8·3	8·1	7·1	2·6	2	7	2	5	4	7	3	5
		×1·5		1·51		35·8	1·11	52	2·28	2·95	1·30	7·7	4·3	10·0	10·5	8·4	2·7	3	9	3	7	6	11	7	12
		×2·0		1·49		39·9	1·44	57	2·58	3·32	1·29	10·3	4·6	12·1	10·3	7·8	2·6	3	10	4	12	9	16	6	10
H	0·30 g	×1·0	1·26	1·43	27	26·3	0·75	46	1·40	1·85	1·32	6·8	4·0	9·7	9·0	6·6	2·4	2	5	2	4	3	5	3	5
		×1·5		1·60		33·5	1·10	52	1·96	2·94	1·50	8·5	4·3	11·8	10·1	6·5	1·7	3	11	3	8	7	11	5	7
		×2·0		1·88		35·3	1·40	57	2·84	4·14	1·46	9·8	5·6	11·4	10·3	6·6	2·1	4	11	5	15	9	15	6	11

Table 5.12. Inelastic response results of 4-storey structures (average for four motions) — 'full slab width' — full capacity design, equation (2.12)

DC	a_g	Exc.	Period: sec		V_b/W: %		$\sqrt{(E_{in}/M)}$:	E_h/E_{in}:	Drift ratio: %			$E_{h,col}/E_{in}$: %		$E_{h,beams}/E_{in}$: %				Damage index: %, storey mean/max.							
																		Columns				Beams			
			T_{cr}	T_{nl}	Ult.	NL	m/s	%	Top	Inter.	Int./Top	1st	Rest/avg.	1st	2nd	3rd	4th	1st		2nd		1st		2nd	
Direction X																									
L	0·15 g	×1·0	1·21	1·24	32	14·4	0·39	45	0·72	0·89	1·23	7·6	6·5	5·4	4·6	3·5	4·6	1	3	1	4	1	4	1	4
		×1·5		1·25		20·8	0·58	42	1·09	1·34	1·24	6·6	5·3	6·9	5·8	4·5	1·9	3	6	3	7	3	6	3	6
		×2·0		1·30		26·7	0·76	42	1·45	1·83	1·26	6·3	4·6	8·4	6·9	5·2	1·2	4	11	4	10	4	8	5	9
M	0·15 g	×1·0	1·18	1·21	33	15·5	0·40	51	0·67	0·88	1·30	7·3	6·2	8·6	6·9	5·0	4·4	1	2	1	3	1	2	1	2
		×1·5		1·24		21·8	0·57	46	1·02	1·36	1·34	6·4	4·9	10·3	8·0	4·7	2·3	2	5	2	5	2	4	2	4
		×2·0		1·30		27·1	0·75	45	1·37	1·84	1·34	6·6	4·1	10·7	8·7	5·1	1·6	3	8	3	7	4	7	4	7
M	0·30 g	×1·0	0·98	1·01	43	33·6	0·82	44	1·15	1·57	1·36	6·5	4·8	7·8	8·0	5·4	1·5	2	5	2	6	3	8	3	7
		×1·5		1·08		41·7	1·18	46	1·71	2·22	1·29	6·5	5·2	7·6	9·0	6·0	1·2	3	7	4	7	5	13	7	16
		×2·0		1·15		50·6	1·51	49	2·35	3·02	1·29	7·1	6·4	7·3	9·4	5·5	0·9	5	13	7	13	9	22	11	26
H	0·30 g	×1·0	1·05	1·10	41	28·9	0·79	44	1·19	1·59	1·34	6·8	4·9	8·2	8·3	4·8	1·4	2	5	2	4	3	5	3	6
		×1·5		1·16		40·8	1·14	46	1·82	2·52	1·39	7·4	4·9	9·2	9·1	5·0	1·0	4	11	4	11	6	12	6	13
		×2·0		1·29		46·1	1·46	52	2·49	3·71	1·49	8·7	6·9	8·9	8·7	4·4	0·7	6	16	8	24	8	16	8	23
Direction Y																									
L	0·15 g	×1·0	1·33	1·37	28	13·4	0·38	48	0·79	0·98	1·23	6·2	4·5	7·4	6·0	8·3	6·5	1	2	1	2	1	2	1	2
		×1·5		1·41		19·0	0·57	45	1·11	1·40	1·27	6·2	4·1	8·3	6·7	7·5	4·2	2	5	2	4	2	4	2	4
		×2·0		1·52		24·5	0·75	45	1·39	1·86	1·34	6·7	3·7	9·3	7·5	6·9	3·5	3	9	2	5	5	8	4	7
M	0·15 g	×1·0	1·32	1·40	26	12·3	0·38	55	0·68	0·98	1·45	7·2	4·1	12·0	9·4	8·0	5·9	1	2	1	1	1	2	1	2
		×1·5		1·52		16·7	0·56	51	1·00	1·49	1·49	7·8	3·7	11·1	9·4	7·3	3·8	1	4	1	2	3	4	2	3
		×2·0		1·56		21·5	0·74	49	1·39	2·04	1·47	8·3	3·5	10·5	9·2	7·1	3·3	2	7	2	4	5	8	4	6
M	0·30 g	×1·0	1·05	1·10	34	28·4	0·79	45	1·22	1·56	1·28	6·6	4·3	7·0	8·5	7·9	2·5	1	3	2	3	3	5	3	5
		×1·5		1·17		39·0	1·13	48	1·88	2·38	1·27	7·2	4·6	8·7	9·3	7·3	2·1	3	9	3	6	8	11	9	13
		×2·0		1·31		46·5	1·46	54	2·69	3·43	1·27	7·3	5·3	9·1	10·9	8·5	2·4	4	11	5	9	14	23	18	27
H	0·30 g	×1·0	1·13	1·24	30	27·0	0·75	47	1·33	1·76	1·32	7·6	3·5	10·4	9·7	6·8	2·1	2	5	2	3	4	5	5	6
		×1·5		1·39		37·4	1·09	48	1·99	2·69	1·35	7·8	3·4	10·7	10·4	7·1	1·8	2	8	3	5	7	10	10	15
		×2·0		1·54		42·5	1·44	52	2·59	3·70	1·43	8·6	4·0	11·4	11·2	7·1	1·7	4	13	4	9	10	15	11	16

Table 5.13. Inelastic response results of 12-storey structures (average for four motions) — 'full slab width' — relaxed capacity design, equation (2.14)

DC	a_g	Exc.	Period: sec		V_b/W: %		$\sqrt{(E_{in}/M)}$:	E_h/E_{in}:	Drift ratio: %			$E_{h,col}/E_{in}$: %		$E_{h,beams}/E_{in}$: %				Damage index: %, storey mean/max. Columns				Beams			
			T_{cr}	T_{nl}	Ult.	NL	m/s	%	Top	Inter.	Int./Top	1st	Rest/avg.	1st	2nd	3rd	4th	1st		2nd		1st		2nd	
Direction X																									
L	0·15 g	×1·0	1·89	2·03	23	12·0	0·35	59	0·45	0·64	1·42	3·8	1·5	4·1	4·5	4·0	3·0	1	2	1	1	2	4	2	4
		×1·5		2·07		17·1	0·53	63	0·66	0·99	1·50	3·8	1·4	4·8	5·4	4·5	3·3	2	5	1	3	5	8	5	9
		×2·0		2·10		20·5	0·72	68	0·86	1·32	1·54	3·9	1·3	5·3	6·8	5·6	3·6	2	8	2	4	8	14	10	17
M	0·15 g	×1·0	2·00	2·17	23	10·8	0·36	66	0·44	0·71	1·62	4·5	1·3	5·5	6·7	5·8	3·3	1	2	0	1	2	4	3	5
		×1·5		2·41		14·0	0·52	67	0·64	1·07	1·65	4·6	1·1	5·9	7·7	6·6	3·4	1	4	1	1	5	8	6	10
		×2·0		2·42		16·6	0·65	70	0·86	1·44	1·68	4·0	1·0	5·5	8·1	6·9	3·8	2	6	1	2	7	13	10	17
M	0·30 g	×1·0	1·72	1·93	32	21·9	0·69	62	0·87	1·18	1·37	3·2	1·4	3·3	4·1	3·8	3·5	2	5	2	4	5	8	6	10
		×1·5		2·02		28·5	1·07	69	1·21	1·69	1·40	3·1	1·2	4·0	5·8	5·1	4·3	3	9	3	6	10	17	13	21
		×2·0		2·10		31·9	1·42	75	1·54	2·27	1·47	3·3	1·0	4·4	7·8	7·6	4·5	4	11	4	9	13	24	21	38
H	0·30 g	×1·0	1·80	2·05	27	20·9	0·72	67	0·84	1·26	1·51	3·7	1·3	5·0	6·5	5·5	3·5	2	4	1	3	6	11	7	12
		×1·5		2·12		25·3	1·06	73	1·18	1·87	1·58	3·8	1·0	5·5	8·9	7·6	3·9	3	8	2	4	11	19	14	24
		×2·0		2·41		27·5	1·32	77	1·58	2·56	1·62	4·3	0·9	6·2	9·9	8·9	4·3	4	10	3	6	15	26	19	32
Direction Y																									
L	0·15 g	×1·0	2·08	2·14	20	11·8	0·37	59	0·51	0·74	1·44	4·5	1·5	4·5	4·9	4·0	2·8	1	3	1	2	3	5	3	5
		×1·5		2·18		15·2	0·53	64	0·74	1·05	1·43	4·2	1·2	4·8	5·6	4·8	3·4	2	5	1	3	6	9	6	9
		×2·0		2·39		17·2	0·67	69	0·96	1·42	1·47	3·9	1·1	5·1	6·7	5·7	4·0	2	6	2	4	8	12	9	14
M	0·15 g	×1·0	2·23	2·48	15	9·7	0·33	62	0·52	0·78	1·52	4·0	1·1	4·8	5·4	4·6	3 5	1	2	0	1	3	4	3	4
		×1·5		2·67		12·0	0·48	70	0·72	1·12	1·56	4·0	1·1	5·4	6·8	5·6	4·1	1	3	1	1	4	6	5	7
		×2·0		2·72		13·4	0·63	74	0·87	1·41	1·62	3·9	1·1	5·0	7·3	6·6	4·4	2	5	1	2	6	9	8	11
M	0·30 g	×1·0	1·85	2·04	26	21·4	0·73	64	0·92	1·30	1·40	3·2	1·3	3·7	4·8	3·5	3·9	2	5	2	4	6	9	6	9
		×1·5		2·13		25·3	1·07	72	1·27	1·91	1·50	3·3	1·1	4·2	6·5	5·6	4·5	3	8	3	6	10	15	12	17
		×2·0		2·18		28·5	1·35	77	1·69	2·32	1·37	3·6	1·0	4·7	7·6	7·3	4·8	4	12	3	8	14	20	17	23
H	0·30 g	×1·0	1·96	2·15	21	18·0	0·70	67	0·92	1·33	1·45	3·7	1·2	4·8	6·4	5·8	3·7	1	4	1	3	6	9	6	10
		×1·5		2·12		21·4	0·98	76	1·28	1·96	1·52	3·6	1·0	5·4	8·3	7·1	4·5	2	7	2	4	10	14	11	16
		×2·0		2·93		23·5	1·27	80	1·60	2·50	1·56	3·5	1·0	5·2	8·9	8·4	4·8	3	9	2	6	14	19	16	24

Table 5.14. *Inelastic response results of 12-storey structures (average for four motions) — 'full slab width' full capacity design, equation (2.12)*

DC	a_g	Exc.	Period: sec		V_b/W: %		$\sqrt{(E_{in}/M)}$:	E_h/E_{in}:	Drift ratio: %			$E_{h,col}/E_{in}$: %		$E_{h,beams}/E_{in}$: %				Damage index: %, storey mean/max. Columns				Damage index: %, storey mean/max. Beams			
			T_{cr}	T_{nl}	Ult.	NL	m/s	%	Top	Inter.	Int./Top	1st	Rest/avg.	1st	2nd	3rd	4th	1st		2nd		1st		2nd	
Direction X																									
L	1·15 g	×1·0	1·89	2·03	23	12·0	0·35	59	0·45	0·64	1·42	3·8	1·5	4·1	4·5	4·0	3·0	1	2	1	1	2	4	2	4
		×1·5		2·07		17·1	0·53	63	0·66	0·99	1·50	3·8	1·4	4·8	5·4	4·5	3·3	2	5	1	3	5	8	5	9
		×2·0		2·10		20·5	0·72	68	0·86	1·32	1·54	3·9	1·3	5·3	6·8	5·6	3·6	2	8	2	4	8	14	10	17
M	0·15 g	×1·0	2·00	2·17	23	10·8	0·36	66	0·44	0·71	1·62	4·5	1·3	5·5	6·7	5·8	3·3	1	2	0	1	2	4	3	5
		×1·5		2·41		14·0	0·52	67	0·64	1·07	1·65	4·6	1·1	5·9	7·7	6·6	3·4	1	4	1	1	5	8	6	10
		×2·0		2·42		16·6	0·65	70	0·86	1·44	1·68	4·0	1·0	5·5	8·1	6·9	3·8	2	6	1	2	7	13	10	17
M	0·30 g	×1·0	1·65	1·88	31	22·3	0·69	63	0·84	1·19	1·41	3·3	1·4	3·6	4·4	4·1	3·5	2	5	2	3	5	9	6	10
		×1·5		1·97		28·8	1·06	69	1·20	1·76	1·46	3·1	1·1	4·0	5·6	5·0	4·4	3	9	3	6	10	17	13	21
		×2·0		2·06		32·9	1·43	75	1·52	2·25	1·48	3·2	1·0	4·4	7·7	7·4	4·6	4	11	4	9	15	28	23	40
H	0·30 g	×1·0	1·72	2·02	25	21·0	0·71	68	0·83	1·23	1·49	3·8	1·3	5·1	6·5	5·5	3·6	2	4	1	3	6	11	7	13
		×1·5		2·12		25·7	1·06	72	1·18	1·92	1·63	4·0	1·0	5·8	9·1	7·7	3·8	3	8	2	4	12	22	15	27
		×2·0		2·42		28·0	1·32	76	1·57	2·59	1·65	4·7	0·9	6·3	9·7	8·8	4·2	4	11	3	5	19	33	23	41
Direction Y																									
L	0·15g	×1·0	2·08	2·14	20	11·8	0·37	59	0·51	0·74	1·44	4·5	1·5	4·5	4·9	4·0	2·8	1	3	1	2	3	5	3	5
		×1·5		2·18		15·2	0·53	64	0·74	1·05	1·43	4·2	1·2	4·8	5·6	4·8	3·4	2	5	1	3	6	9	6	9
		×2·0		2·39		17·2	0·67	69	0·96	1·42	1·47	3·9	1·1	5·1	6·7	5·7	4·0	2	6	2	4	8	12	9	14
M	0·15 g	×1·0	2·23	2·48	15	9·7	0·33	62	0·52	0·78	1·52	4·0	1·1	4·8	5·4	4·6	3·5	1	2	0	1	3	4	3	4
		×1·5		2·67		12·0	0·48	70	0·72	1·12	1·56	4·0	1·1	5·4	6·8	5·6	4·1	1	3	1	1	4	6	5	7
		×2·0		2·72		13·4	0·63	74	0·87	1·41	1·62	3·9	1·1	5·0	7·3	6·6	4·4	2	5	1	2	6	9	8	11
M	0·30 g	×1·0	1·80	2·01	25	21·9	0·72	63	0·92	1·31	1·43	3·4	1·3	3·8	4·8	3·5	3·8	2	5	1	3	6	9	7	9
		×1·5		2·10		26·6	1·09	70	1·29	2·03	1·57	3·4	1·0	4·2	6·7	6·1	4·3	3	8	2	5	12	16	15	19
		×2·0		2·18		30·0	1·36	76	1·68	2·37	1·41	3·4	1·0	4·5	7·8	7·1	4·8	4	12	3	7	17	23	22	30
H	0·30 g	×1·0	2·00	2·18	20	18·9	0·71	65	0·91	1·31	1·45	3·9	1·1	5·1	6·5	5·6	3·5	2	5	1	3	7	11	8	11
		×1·5		2·40		22·8	0·98	74	1·34	2·00	1·50	3·7	1·0	5·5	8·4	7·2	4·3	3	8	2	4	13	18	15	22
		×2·0		2·65		24·5	1·28	79	1·76	2·71	1·54	3·7	1·0	5·1	9·2	8·7	4·6	4	10	2	5	18	26	24	34

Table 5.15. Inelastic response results of 3-storey structures (average for four motions) — 'full slab width' — relaxed capacity design, equation (2.14)

DC	a_g	Exc.	Period: sec		V_b/W: %		$\sqrt{(E_{in}/M)}$:	E_h/E_{in}:	Drift ratio: %			$E_{h.col}/E_{in}$: %		$E_{h,beams}/E_{in}$: %		Damage index: %, storey mean/max.							
																Columns				Beams			
			T_{cr}	T_{nl}	Ult.	NL	m/s	%	Top	Inter.	Int./Top	1st	Rest/avg.	1st	2nd	1st		2nd		1st		2nd	
Direction X																							
L	0·15 g	×1·0	1·94	2·00	20	9·8	0·31	16	1·04	2·03	1·95	10·8	1·8	1·7	0·5	4	6	1	2	1	3	0	1
		×1·5		2·04		14·2	0·46	21	1·56	3·04	1·94	12·2	1·6	4·9	0·6	8	12	2	3	2	7	1	2
		×2·0		2·07		15·4	0·63	39	2·11	4·27	2·03	30·0	1·2	6·4	0·6	11	19	2	5	3	11	1	3
M	0·15 g	×1·0	1·87	2·00	20	9·9	0·32	21	1·04	1·92	1·85	11·2	2·1	5·3	0·6	3	4	1	2	1	3	0	1
		×1·5		2·05		13·4	0·47	33	1·54	2·83	1·83	16·0	1·7	12·9	0·6	5	7	2	3	3	8	1	2
		×2·0		2·07		15·4	0·62	40	2·05	3·87	1·89	23·0	1·5	13·5	0·7	8	11	3	5	6	15	1	3
M	0·30 g	×1·0	1·51	1·61	28	26·2	0·70	43	1·42	2·49	1·76	21·7	2·6	14·5	1·7	6	10	2	3	6	13	2	4
		×1·5		1·81		30·1	0·99	55	2·03	3·76	1·85	32·9	2·0	15·6	2·1	8	14	3	5	11	22	3	6
		×2·0		1·89		31·9	1·32	62	2·67	5·40	2·02	42·3	1·9	13·3	2·1	15	25	4	6	13	29	4	9
H	0·30 g	×1·0	1·57	1·84	25	21·5	0·66	32	1·29	2·21	1·72	12·4	2·0	14·4	1·2	3	6	1	2	6	9	1	3
		×1·5		1·88		27·0	0·99	46	1·96	3·44	1·76	23·6	1·6	16·9	1·8	6	9	2	3	12	22	3	5
		×2·0		2·21		28·7	1·32	50	2·54	4·95	1·95	31·2	1·5	14·6	1·6	9	15	2	4	14	27	3	7
Direction Y																							
L	0·15 g	×1·0	2·14	2·15	23	13·5	0·33	23	0·80	1·28	1·60	12·1	3·4	2·2	1·9	1	3	1	2	1	3	1	3
		×1·5		2·17		19·1	0·52	28	1·17	1·91	1·64	15·2	3·4	5·3	2·4	2	6	2	5	3	6	3	5
		×2·0		2·17		21·8	0·69	53	1·48	2·52	1·70	36·4	3·9	8·1	3·8	4	11	3	7	6	10	5	9
M	0·15 g	×1·0	2·09	2·10	23	13·5	0·33	23	0·80	1·28	1·60	12·1	3·4	2·2	2·0	1	2	1	1	1	2	1	2
		×1·5		2·10		19·1	0·49	28	1·19	1·95	1·64	15·1	3·1	5·0	1·8	2	4	1	3	3	5	3	4
		×2·0		2·13		21·9	0·66	50	1·49	2·72	1·83	35·1	2·5	8·0	2·1	3	7	2	5	5	8	5	7
M	0·30 g	×1·0	1·06	1·16	43	30·3	0·78	35	1·07	1·40	1·30	11·7	3·0	13·6	3·3	2	6	2	4	4	7	3	6
		×1·5		1·24		40·0	1·13	48	1·70	2·09	1·23	16·1	2·7	19·7	7·1	4	9	3	9	7	12	6	10
		×2·0		1·26		46·8	1·48	63	2·88	3·42	1·19	22·0	6·7	18·3	9·1	5	14	5	13	11	19	8	13
H	0·30 g	×1·0	1·13	1·25	37	29.1	0·76	37	1·12	1·56	1·39	12·0	2·7	16·5	2·7	2	6	2	5	4	6	4	6
		×1·5		1·40		36·8	1·10	47	1·47	2·40	1·38	17·7	2·1	19·0	5·7	4	11	4	9	8	12	6	10
		×2·0		1·65		41·4	1·45	51	2·22	3·29	1·48	23·2	2·7	16·7	5·6	5	12	5	13	8	14	7	12

Table 5.16. Inelastic response results of 3-storey structures (average for four motions) — 'full slab width' — full capacity design, equation (2.12)

DC	a_g	Exc.	Period: sec		V_b/W: %		$\sqrt{(E_{in}/M)}$:	E_h/E_{in}:	Drift ratio: %			$E_{h,col}/E_{in}$: %		$E_{h,beams}/E_{in}$: %		Damage index: %, storey mean/max. Columns				Beams			
			T_{cr}	T_{nl}	Ult.	NL	m/s	%	Top	Inter.	Int./Top	1st	Rest/avg.	1st	2nd	1st		2nd		1st		2nd	
Direction X																							
L	0·15 g	×1·0	1·94	2·00	20	9·8	0·31	16	1·04	2·03	1·95	10·8	1·8	1·7	0·5	4	6	1	2	1	3	0	1
		×1·5		2·04		14·2	0·46	21	1·56	3·04	1·94	12·2	1·6	4·9	0·6	8	12	2	3	2	7	1	2
		×2·0		2·07		15·4	0·63	39	2·11	4·27	2·03	30·0	1·2	6·4	0·6	11	19	2	5	3	11	1	3
M	0·15 g	×1·0	1·66	1·69	19	11·6	0·32	20	0·74	1·10	1·49	0·0	3·0	12·3	1·6	0	0	1	1	2	4	1	1
		×1·5		1·82		16·9	0·48	27	1·11	1·67	1·51	0·0	3·0	17·8	3·0	0	0	1	2	6	9	2	3
		×2·0		1·85		22·0	0·65	30	1·52	2·30	1·51	0·0	3·0	19·5	4·9	0	0	2	3	10	17	3	7
M	0·30 g	×1·0	1·49	1·56	33	28·2	0·70	35	1·38	2·49	1·80	14·5	2·5	13·9	1·9	6	8	1	2	7	15	2	4
		×1·5		1·64		33·2	1·01	47	1·92	3·49	1·82	19·3	1·9	20·9	3·0	10	14	2	3	15	28	5	7
		×2·0		1·85		33·8	1·31	56	2·57	5·07	1·97	39·8	1·5	19·4	2·8	14	17	3	5	24	44	7	11
H	0·30 g	×1·0	1·51	1·59	34	24·4	0·68	31	1·26	2·26	1·80	11·0	1·9	14·1	1·6	4	7	1	2	6	11	2	3
		×1·5		1·67		31·2	1·00	45	1·84	3·28	1·78	14·8	1·7	22·9	3·2	7	12	2	3	18	27	4	8
		×2·0		1·86		32·9	1·32	49	2·50	4·57	1·83	19·1	1·5	23·5	3·7	10	15	3	5	29	44	7	12
Direction Y																							
L	0·15 g	×1·0	2·14	2·15	23	13·5	0·33	23	0·80	1·28	1·60	12·1	3·4	2·2	1·9	4	7	1	2	0	1	0	0
		×1·5		2·17		19·1	0·52	28	1·17	1·91	1·64	15·2	3·4	5·3	2·4	8	14	2	3	1	1	0	0
		×2·0		2·17		21·8	0·69	53	1·48	2·52	1·70	36·4	3·9	8·1	3·8	11	21	2	4	1	3	0	1
M	0·15 g	×1·0	1·90	1·96	22	10·9	0·33	5	0·80	1·68	2·11	0·0	1·0	1·8	0·6	0	0	0	0	0	1	0	0
		×1·5		1·96		16·3	0·49	5	1·19	2·51	2·10	0·0	1·0	2·5	0·6	0	0	0	1	1	2	0	0
		×2·0		1·99		20·8	0·67	10	1·58	3·30	2·08	0·0	1·1	7·4	0·6	0	0	1	2	2	4	0	1
M	0·30 g	×1·0	1·02	1·08	50	32·8	0·78	32	1·01	1·49	1·48	11·4	2·1	12·9	3·3	3	7	1	2	3	5	1	2
		×1·5		1·14		44·1	1·16	38	1·59	2·17	1·37	13·0	2·1	14·6	6·3	6	14	2	3	7	10	3	6
		×2·0		1·19		48·9	1·50	45	2·22	3·09	1·39	18·9	1·7	15·3	7·2	10	23	2	5	11	17	6	10
H	0·30 g	×1·0	1·10	1·17	45	31·5	0·78	33	1·03	1·48	1·44	11·8	2·1	13·7	3·6	3	7	1	2	3	5	1	2
		×1·5		1·21		42·7	1·15	41	1·63	2·19	1·34	13·6	2·0	15·4	7·5	6	14	1	3	7	10	3	6
		×2·0		1·27		47·2	1·38	49	2·31	3·09	1·34	18·0	1·7	18·0	ty -39·3		9	20	2	5	11	16	6

predicted in design, i.e. typically in the second storey of the 4-storey structures and always in the first of the 3-storey ones. However, in the 12-storey structures the maximum inter-storey drift takes place anywhere between storeys 2 and 11, depending on the details of the structure and of the ground motion, whereas in the design of these buildings the maximum inter-storey drift was invariably in the 3rd storey. This shows the increased importance of higher modes in the response of the 12-storey structures, due to the lengthening of their periods. The shift of maximum inter-storey drifts to the upper storeys is more pronounced in the structures considered with zero participating slab width: in the 12-storey structures, this maximum occurs more often in storey 9 or 10 than in storey 3 or 4. The reason lies not only in the aforementioned increased importance of the second or third modes, but also in the fact that in the lighter reinforced upper storey beams, the inclusion or disregard of the contribution of slab reinforcement to the tension flange has a larger impact on the stiffness of the corresponding storey.

Peak response top and inter-storey drift ratios are overestimated on average by 25% and 15% respectively, by the rule of equal displacements of the 5%-damped elastic and inelastic structure, with the members of the elastic structure considered at incipient yielding. The top and inter-storey drift ratios of the structures with columns proportioned according to full capacity design, equation (2.12), are lower than those of the structures following the relaxed capacity design, equation (2.14). The reason is the larger cracked section stiffness and the less pronounced inelasticity of the more heavily reinforced columns. When the contribution of slab reinforcement to the beam tension flange is neglected, computed peak top and maximum inter-storey drifts generally increase.

The ratio of maximum to average inter-storey drift in the structure, listed in the 12th column of Tables 5.11–5.16, is a measure of the real tendency for soft-storey development. With the exception of the 3-storey structures, at intensity 1·0 this ratio is about the same as in the design calculations. In the full slab width structures subjected to the relaxed capacity design of equation (2.14), the ratio of the maximum to the average inter-storey drift typically increases slightly with the increase in the excitation intensity from 1·0 to 2·0. On the contrary, in the structures which follow the full capacity design of equation (2.12) and in those considered with zero slab width participation on the beams, this ratio does not always increase with motion intensity, and may even decrease. In the structures considered with full slab width and with columns proportioned according to the relaxed capacity design of equation (2.14), the increase in the ratio of maximum to average inter-storey drift with increasing intensity of excitation is larger in DCM than in DCL structures designed for a ground acceleration of 0·15 g, and in DCH than in DCM ones designed for 0·3 g. This effect is not noticeable, however, when the columns are proportioned according to the full capacity design, equation (2.12), especially when the beams are considered as in design, i.e. with zero effective slab width in tension. This means that the stricter capacity design provisions of Eurocode 8 for higher DC structures are effective in preventing the tendency towards soft-storey formation, especially when full capacity design is employed. However the unavoidable participation of slab reinforcement in beam bending destroys the intended column–beam strength balance. As a result the increased inelasticity of structures designed for lower strength and higher ductility, inherently tends to develop through storey-sway mechanisms rather than beam-sway ones. In the lower rise 3-storey structures, the ratio of maximum to average inter-storey drifts is very large, which is consistent with the very large drift ratios of the first storey of these structures. The reduction of the ratio of the maximum to the average inter-storey drift ratio, especially at higher motion intensities, is the clearest beneficial effect of the full capacity design of columns according to equation (2.12), as it suggests a reduction in the tendency for formation of a storey-sway mechanism demonstrated by the structures which follow the relaxed capacity design of equation (2.14). Another interesting result is that when the slab participation in tension is considered as in design, i.e. practically zero, the ratio of the maximum inter-storey to top drift ratios is, in general, lower than when the slab participation is considered as in reality, i.e. almost full, suggesting a smaller tendency of the as-designed structure towards soft-storey development. The 4-storey structures are an exception to this rule, as their inter-storey drifts are disproportionately increased in the middle to upper storeys when the slab participation in tension is neglected. Overall, consistent with the lower than 1·0 values of the sway potential index S_p in Figs 5.39 and 5.40, results do not suggest any tendency towards soft-storey formation.

The large drift ratios imply very significant P–δ effects, which are taken into account in the non-linear analyses. The magnitude of these effects, relative to the first-order ones of the seismic action, can be estimated as the inter-storey drift ratio divided by the peak response base shear coefficient V_b/W given in the 7th column of Tables 5.11–5.16.

In the structures which follow the relaxed capacity design of equation (2.14), the peak base shear coefficients developed during the response to the design intensity motion are about equal to the design base shear coefficient for the DCL structures, but exceed them by as much as 50% for DCM structures, or by 50–100% for DCH ones. At stronger intensity motions, base shears increase roughly in proportion to the motion intensity, with a tendency to saturate to 2·0–2·5 times the design base shear for DCH and DCM structures at an excitation with twice the design intensity. The peak base shear coefficients during the response are higher in the structures with full capacity design, equation (2.12), than in those with relaxed, equation (2.14). When a zero slab width in tension is considered, peak response base shear coefficients in general decrease over the corresponding values for full slab width participation, due to the reduction in stiffness and the lengthening of the predominant periods of vibration.

As a result of strain hardening of the members after yielding, peak response base shears are sometimes higher than the values listed in the 6th column of Tables 5.11–5.16 as 'ultimate' base shears. These values correspond to a full or partial beam sidesway mechanism, under gradually increasing static lateral storey forces proportional to the product of the storey mass and its height from the ground, and have been computed through an incremental non-linear 'pushover' analysis. Both the 'ultimate' and the peak response base shears are less affected by ductility class than the design base shears. The reason is that these values reflect the real resistance of the structure, as this is affected by the various sources of overstrength on the supply side. Many of these sources, such as the control of cross-sectional dimensions by drift limitations and the control of reinforcement by gravity loads or by minimum requirements, are independent of the ductility class.

The difference between the ultimate or the peak response base shear coefficient and the corresponding design value reflects the available overstrength on the supply side. The most systematic source of overstrength is the difference between the mean strengths of steel and concrete, on which the non-linear analyses are based, and the design values of these strengths. The ratios of the former to the latter are in this case equal to about 1·3 and 2·0 for steel and concrete respectively, values which are typical of actual conditions in practice. Other systematic sources of overstrength are: the rounding-up of bars during proportioning of the reinforcement, the control of the required steel area by minimum measures and detailing rules, by gravity loads, or by neighbouring cross-sections (as in the two beam or column sections on opposite sides of a joint covered by the same reinforcement), the capacity design magnification of column moments (especially for structures with columns following the full capacity design of equation (2.12)), the contribution of slab reinforcement to the strength of the beam top flange in tension, when this contribution is accounted for, etc. Another source of overstrength, which is, however, present only in this case of non-linear response analysis to unidirectional ground motions without accidental eccentricity, is the fact that beams and columns have been designed for the effects of an accidental eccentricity of 5% of the dimensions of the structure in plan and for a simultaneous orthogonal component equal to 30% of the design seismic action. In the present case of beams running parallel to the X and Y directions of application of the seismic action, beam overstrength due to these effects amounts only to 6% and is due to the accidental eccentricity alone. However, for the columns these two sources of overstrength may increase resistance by more than 50%. If the non-linear analyses were performed for simultaneous excitation in the two orthogonal horizontal directions and with eccentricities of the structural masses as considered in design, then this important source of overstrength would not have been present.

The various overstrengths on the supply side accumulate to an aggregate overstrength factor ranging in the present case from about 1·4 in some members (mainly the lower storey beams) to more than 2·0 in others (mainly in the columns, especially those of the upper storeys). The effect of this overstrength on the response is combined with that of the demand-side understress due to softening of the structure: structures are required to develop inelastic action equivalent to their design q-factor multiplied by the reduction of the elastic response due to softening, i.e. in the present case by about 0·55 for 0·15 g structures or about 0·6 for 0·3 g ones subjected to the design intensity motion, and divided by the supply-side overstrength factor. For DCL structures the net result may be a q-factor of 1·0 or less, which means that their response to the design intensity motion may be elastic.

Tables 5.11–5.16 present, in the 9th column, the total energy absorbed by hysteresis by the end of the 10 s of the non-linear analysis E_h as a percentage of the total input energy E_{in}. The ratio of E_{in} to the total mass of the structure, given in the 8th column of the Tables, is approximately equal to the square of the pseudovelocity in the constant velocity range of the excitation spectrum, where the period of the cracked elastic structure lies. Owing to their higher strength, the structures following the full capacity design of equation (2.12) have a slightly higher total energy input than the ones with the relaxed capacity design of equation (2.14). However, the effect of either considering or disregarding the participation of the slab in tension is smaller and less systematic: typically, but not always, the total energy input increases slightly when the slab participation is considered. For the same structure the ratio E_h/E_{in} increases little with input motion intensity, with the exception of the 3-storey structure, in which the energy absorption increases dramatically when the excitation intensity doubles. In structures following the full capacity design of equation (2.12), the hysteretic–total input energy ratio is noticeably lower than in those following the relaxed capacity design of equation (2.14), and increases even less with motion intensity, demonstrating the less pronounced overall inelastic action of these structures with the stronger columns. Similarly, the value of E_h/E_{in} is higher when the contribution of the slab to the tension flange of the beam is neglected, as the reduced flexural capacity of the beams in negative bending increases the overall extent of the non-linearity of the response. The 0·15 g 3-storey structures are an exception to this rule, as in them the reduced flexural capacity of the beams without the slab width participation fully prevents plastic hinging in the first storey columns and hysteretic energy dissipation there. For the same design acceleration the energy absorption ratio slightly increases with increasing DC. In general, however, as E_h/E_{in} is a measure of the non-linearity of the response, the more control over the design exerted by the seismic action, the higher is this ratio for the same motion intensity. So, within the same class of structures, E_h/E_{in} is higher in the 0·30 g structures than in the 0·15 g ones. This effect is very clear in the 3-storey structures, which, at least for the 0·15 g design acceleration, are governed by gravity loads and not by the seismic action. At the other extreme, in the 12-storey structures, the designs of which are totally controlled by the seismic action, hysteretic energy dissipation is the highest.

Columns 13 to 18 of Tables 5.11 to 5.14 and 13 to 16 of Tables 5.15 and 5.16, give the breakdown of the hysteretic energy dissipation among the members and the storeys of the structure. A clear effect in these results is that, due to capacity design, increasing the DC for the same design acceleration noticeably increases the contribution of beams in the energy dissipation. This is more so in the structures with the full capacity design of columns. Moreover, in general the total energy dissipation in the columns is typically less, and indeed significantly less, than in the beams. The exception that confirms the rule occurs in some of the 3-storey structures with columns designed according to equation (2.14) rather than to equation (2.12), in which capacity design of columns in flexure is most relaxed due to the controlling gravity loads.

An interesting observation is that the share of columns in the hysteretic energy dissipation follows the value of the sway potential index S_p, being larger when S_p is also larger (e.g. in the 4-storey structures it is larger in the X direction than in the Y and vice-versa for the 3-storey ones, while in the 12-storey structures, which have very low S_p values, columns contribute very little to the energy dissipation).

As expected, the fraction of input energy dissipated in columns is lower in the structures with full capacity design according to equation (2.12) than in those following the relaxed capacity design format of equation (2.14), and slightly lower in those in which the participation of the slab to the tension flange of the beams is neglected than in those in which it is considered. Also, in the former class of structures this fraction increases less with input motion intensity than in the latter, suggesting limited inelastic action in columns even at higher intensities. The first storey columns of all 3-storey 0·15 g structures in which the slab contribution in tension is neglected, and of the 0·15 g DCM ones designed with full capacity design and considered with 'full slab width' remain elastic for both directions of the seismic action and at all motion intensities. This is surprising for the Y direction of the DCL 3-storey structures, as their sway potential index S_p exceeds 1·0 (Figs 5.39 and 5.40). This means that for the 0·15 g 3-storey structures with the beam reinforcement as considered in design (i.e. without the slab contribution), capacity design is not essential to protect the ground-storey columns against plastic hinging. When the beam overstrength due to the slab contribution in tension is considered, protection of these columns from plastic hinging is possible only through the large overstrength in both directions of bending, derived by ground-storey columns from their large capacity design magnification required to balance fully the flexural capacity of the long span beams in the Y direction.

The fraction of input energy dissipated in beams is typically (but not always) higher in structures with full capacity design of columns, than in those with relaxed. Overall, it is also considerably higher when the contribution of slab reinforcement to the beam strength is neglected. Moreover, the distribution of energy dissipation among the beams is significantly affected by the consideration or disregard of slab reinforcement: when this reinforcement is neglected, energy absorption is shifted to the beams of the upper storeys, as it is there that the slab reinforcement most increases the beam flexural capacity and most upsets the column–beam strength balance. In other words, the as-intended structures have a rather uniform distribution of energy absorption in the beams of most storeys, rather than a concentration in the lower-storey beams and columns.

Tables 5.11–5.16 list in the last eight columns the average value of the member damage index per storey, separately for beams and columns, as well as the (average over the four accelerograms) maximum beam or column damage value in the storey. Due to space limitations damage values are only given for the first and second storey. Maximum member damage index values well in excess of those listed in Tables 5.13 and 5.14 take place at intermediate and upper storeys of the 12-storey structures. Indeed, in most cases the maximum over the structure damage index value occurs in a 4th-storey to 10th-storey beam.

Storey-average and storey-maximum damage index values have very low values in the columns, and low to sometimes medium values in the beams. In structures with full capacity design, equation (2.12), columns exhibit lower damage index values (especially as far as the maximum column damage in the storey is concerned) than in the structures with relaxed capacity design, equation (2.14). Beam damage is typically lower in the former class of structures than in the latter at low input motion intensities, but the situation is reversed at high intensities. Column damage is also lower in the structures with the slabs as-intended (i.e. with almost zero participation) than in the structures as they really are. The difference is largest when the smaller beam strength of the former case prevents plastic hinge formation in the columns. Neglecting the slab contribution to beam strength significantly increases beam damage in the upper and middle storeys. Designing for a higher DC but for given design ground acceleration, reduces on average the damage index values. As this effect is observed equally in beams and in columns, it is the result of better member detailing, rather than of the more stringent capacity design rules of the higher ductility classes. So a higher DC structure has superior performance in terms of member damage and integrity than a lower DC one, although it develops on average approximately similar displacements and absorbs about the same hysteretic energy. Another clear effect is that the 0·15 g structures suffer significantly less damage than the 0·30 g ones of the same or of different ductility class, when subjected to the same multiple of their design seismic action. Member resistance in the 0·15 g structures is often controlled by gravity loads and by minimum reinforcement, therefore providing more overstrength against the design seismic action.

5.3.5.5. Concluding remarks. Despite the fact that the sway potential index S_p is almost always less than 1·0, and sometimes significantly less, and regardless of whether the contribution of slab reinforcement to the beam top flange is considered or not, inelastic action is not limited to beams but is also spread to columns, as evidenced by the hysteretic energy dissipation and damage therein. This is particularly so if the relaxed capacity design approach of equation (2.14) is followed, instead of the full capacity design of equation (2.12). Nevertheless, even in structures with relaxed capacity design, this inelastic action is limited and does not lead to soft-storey formation or seem to endanger the integrity of the structure. When the beams are considered as intended in the design, i.e. with very little slab participation in their tension flange, inelastic action and damage in the columns is reduced, but, with few exceptions, it is not prevented.

In conclusion, although the relative flexural capacities of beams and columns in a storey suggest that inelastic action will be limited to the beams, some inelastic action does develop in the columns as well. In the structures as considered in design, i.e. practically without slab participation to the tension flange, the distribution of inelasticity and damage is closer to that intended in design, i.e. very limited in the columns and rather uniformly spread to the beams of all storeys. This favourable pattern of inelastic response is not overly distorted by the large beam overstrengths due to the contribution of the slab in negative bending: any column inelastic action that may develop in the as-built structure under very strong ground motions does not seem likely to lead to large inter-storey drifts in a single storey and to formation of a storey-sway mechanism. The conventional, full capacity design of columns in bending significantly improves their protection against inelastic action and plastic hinge formation, without precluding it. Structural damage and inelastic energy dissipation in columns is noticeably reduced by full capacity design, along with any tendency for soft-storey formation. Nevertheless, these favourable effects of full capacity design are much less than expected at first sight on the basis of the much higher moment magnification effected by full capacity design, in comparison to that resulting from the relaxed capacity design of equation (2.14) (cf. Figs 5.28 and 5.29). Capacity design magnification effects are often overshadowed by other sources of overstrengths in the columns, such as that due to minimum reinforcement.

5.4. Conclusions

The case studies reported in section 5.2 on the one hand and 5.3 on the other, follow different routes for the investigation of the effectiveness of different approaches, codified or not, to the control of inelastic seismic response. Those in section 5.2 follow option (*a*) of section 5.3.1, i.e. in which one capacity design parameter is varied at a time, whereas those in section 5.3 follow option (*b*), i.e. they adopt the simultaneous variation of different 'ductility class' parameters, all aiming simultaneously at increasing or reducing the global and local ductility capacity of the structure. Accordingly, the different case studies in these two sections allow, in general, conclusions to be drawn on different aspects of normative or prenormative efforts to control inelastic response. Nevertheless, these case studies provide the ground for some common conclusions as well.

The most important common conclusion of all the case studies in this chapter is that none of the current code approaches to the capacity design of columns in flexure can protect them from plastic hinge development. This is the case for the following

(*a*) the conventional capacity design format of equation (2.12) with $\gamma_{Rd} = 1{\cdot}35$ or $1{\cdot}2$ (for DCH and DCM, respectively) in section 5.3.5, or with $\gamma_{Rd} = 1{\cdot}0$ (with or without the dynamic magnification factor ω of the New Zealand code) or $\gamma_{Rd} = 1{\cdot}5$ in section 5.2;

(*b*) the conventional capacity design format of equation (2.12) with $\gamma_{Rd} = 1{\cdot}4$ and the magnification factor applied only on the seismic moments from the analysis, as in the Greek Code (section 5.2); and

(*c*) the relaxed capacity design procedure of equation (2.14) allowed by Eurocode 8, for DCM, i.e. with $\gamma_{Rd} = 1{\cdot}2$ in section 5.2, or for $\gamma_{Rd} = 1{\cdot}35$ and $\gamma_{Rd} = 1{\cdot}2$ in section 5.3, both when the column overstrength due to the Eurocode 8 detailing provisions is considered (sections 5.2 and 5.3) or neglected (section 5.2).

In all these cases, column plastic hinging takes place in several locations above the ground-storey base, under the design ground motion when column sections are tapered at intermediate storeys and most sources of overstrength are neglected (as in section 5.2), or between 1·5 times and twice the design motion when column sections are kept constant and almost every conceivable source of overstrength is accounted for in the non-linear analysis (as in section 5.3). Nevertheless, column hinge patterns at any instant in time are far from simultaneous plastic hinging at top and bottom of the same storey, and the peak inter-storey drift values attained during the response, as well as their heightwise distribution, do not indicate a tendency for soft-storey formation.

Obviously the extent of plastic hinging and the ductility demands in the columns decrease as their capacity design 'protection' increases (i.e. if the value of γ_{Rd} increases or the dynamic magnification factor ω is introduced, or when the relaxation of capacity design according to equation (2.14) is dropped in favour of the conventional format of equation (2.12)), but column inelastic action is not avoided. Plastic hinging in the columns above the ground-storey base can be avoided, however, at this level of the seismic action if the seismic action effects in the columns are simply doubled. The magnification factor of 2·0, which was found in section 5.2 to be quite effective in preventing column plastic hinging at this seismic action level, is, in general, higher than the average column magnification factor in the designs of the 26 buildings according to Eurocode 8 in section 5.3.2: that factor typically ranges between 1·0 and 1·2, rarely going up to 1·4, when the relaxed version of capacity design, equation (2.14), is applied, or between 1·0 and 2·0, going up to 3·0 in some cases, when the conventional capacity design procedure, equation (2.12), is used in the design. The new procedure proposed in section 5.2 was also found to be quite effective in limiting column plastic hinging only to the ground-storey base, under this level of the seismic action. It should be remembered that in this procedure the column seismic moments to be used for their proportioning are required to envelop the elastic column moments computed in non-linear dynamic analyses of the structural response to an ensemble of ground motion time-histories. In these non-linear analyses, the columns are modelled as elastic, while a non-linear point-hinge model is used for the beams, which have been proportioned previously in flexure for the strength demands of the design seismic action. It is noteworthy that this approach is conceptually similar to that of the AIJ (1990) guidelines reviewed in section 2.5, the difference being that according to those guidelines, instead of a non-linear dynamic analysis, a non-linear static 'push-over' analysis under monotonically increasing lateral loads with an inverted triangular distribution is used for the estimation of the peak moment demands in the elastic columns and a semi-empirical dynamic magnification factor ω is applied to them to account for higher mode effects.

Another common conclusion of the case studies in sections 5.2 and 5.3 is that in dual structures the walls are quite effective in limiting the inter-storey drifts and in preventing not only soft-storey formation, but also in most cases any inelastic action in the columns. Even when such action develops, it leads to very low ductility demands. The relaxation of column capacity design according to equation (2.14), allowed by Eurocode 8, effected in this case due to the low value of the beam moment reversal factor δ, can quite reasonably be applied. Furthermore, since the column magnification factors resulting from the relaxed capacity design in dual structures are only slightly higher than 1·0 (cf. Fig. 5.28), it may make sense, for simplicity, to drop the capacity design requirement for the columns of dual structures, if the strength or stiffness capacity of the walls of such structures exceeds a certain limit. In view of the difficulty and the possible arbitrariness in the selection of this limit, it makes sense to retain the relaxed capacity design requirement for the columns of dual structures, which automatically (yet indirectly, through δ) takes into account the protection offered to columns by the walls.

The case studies in section 5.3 allow some more conclusions which apply specifically to Eurocode 8. First, all these studies show that the three ductility classes of Eurocode 8 provide effectively the same level of control of the inelastic response. Secondly, structures designed to Eurocode 8 seem to possess a significant average overstrength, to cope with uncertainties in the seismic action and demand as well as in the strength and ductility capacities of the members and of the structure as a whole. This overstrength is significantly larger in low to moderate seismicity regions, as there structures are to a large extent gravity-dominated. According to section 5.3.4, due to this overstrength, and almost regardless of their ductility class, structural system and configuration, structures designed for a ground acceleration of 0·15 g or 0·30 g possess a behaviour-factor capacity which exceeds by about 4 or 2·5 times respectively their code-specified behaviour factor. This is consistent with the findings of sections 5.3.4 and 5.3.5 that under twice the design seismic action the most critically stressed members in the structure do not exceed one-quarter to one-third of their ultimate deformation capacity. It is noteworthy that, according to section 5.3, a major part of this overstrength comes from the softening of the structure due to concrete cracking.

One conclusion of section 5.3, which is contrary to the rest as far as the implications for the effectiveness of the Eurocode 8 provisions are concerned, is that structural walls seem to develop significant inelastic action and damage above the base region considered as critical by Eurocode 8. This can be interpreted as a sign that proportioning of the walls in flexure according to the linear moment envelope defined at the base by the wall moment there from the analysis M_{Sw} is not sufficient to protect the regions higher up from significant inelastic action. The New Zealand code provision, according to which the linear moment envelope is defined at the base by the flexural capacity of the wall there, as-detailed and at overstrength, $\gamma_{\mathrm{Rd}}M_{\mathrm{Rw}}$, may be a better alternative.

Appendix 5.1. Comparison of EC8 ductility class requirements

Beams	**Ductility class L**	**Ductility class M**	**Ductility class H**
Design bending moments	From analysis		
Design shear forces	From analysis		Consistent with section design for bending with $\gamma_{Rd} = 1{\cdot}25$
Min. tens. reinf. ratio: %	$50f_{ctm}/f_{yk}$*		
Max. tens. reinf. ratio: %	3	$65f_{cd}/f_{yd}\rho'/\rho + 0{\cdot}15$†	$35f_{cd}/f_{yd}\rho'/\rho + 0{\cdot}15$§
Critical region length	$1{\cdot}0h$	$1{\cdot}5h$	$2{\cdot}0h$
Min. hoop diameter	6 mm		
Max. hoop spacing outside critical region	(i) If $V_{Sd} < V_{Rd2}/5$ min. of ($0{\cdot}8h$, 300 mm) (ii) If $V_{Rd2}/5 < V_{Sd} < 2V_{Rd2}/3$ min of ($0{\cdot}6h$, 300 mm) (iii) If $2V_{Rd2}/3 < V_{Sd}$ min of ($0{\cdot}3h$, 200 mm)		
Max. hoop spacing inside critical region	As above	Min. of ($h/4$, $24\phi_h$, 200 mm, $7\phi_l$)	Min. of ($h/4$, $24\phi_h$, 150 mm, $5\phi_l$)
Shear design (EC2)	(i) If $V_{Sd} < V_{Rd1}$ min. shear reinforcement (ii) If $V_{Sd} > V_{Rd1}$ $V_{Sd} < V_{Rd3} = V_{Rd1} + V_{wd}$ (sum of concrete and stirrup contributions)		
Concrete contribution to shear resistance outside critical region	V_{Rd1}	$0{\cdot}4V_{Rd1}$	0
Concrete contribution to shear resistance inside critical region	As above	(i) If $\zeta > -0{\cdot}5$ (low shear reversal) as for EC2 above (ii) If $\zeta < -0{\cdot}5$ (full shear reversal and) $\lvert V_S \rvert_{max} < X(2+\zeta)\tau_{Rd}bd$	
Shear reinforcement	As for EC2 above	(i) For $X = 4$ as for EC2 above (ii) For $4 < X < 8$ $0{\cdot}5V_{Smax}$ resisted by stirrups and $0{\cdot}5V_{Smax}$ resisted by inclined reinforcement (iii) For $X > 8$ V_{Smax} resisted by bi-diagonal reinforcement	(i) For $X = 3$ as for EC2 above (ii) For $3 < X < 6$ $0{\cdot}5V_{Smax}$ resisted by stirrups and $0{\cdot}5V_{Smax}$ resisted by inclined reinforcement (iii) For $X > 6$ V_{Smax} resisted by bi-diagonal reinforcement

Columns	**Ductility class L**	**Ductility class M**	**Ductility class H**
Bending moments	From analysis	Amplified by $1 + (\alpha_{CD}^{**} - 1)\delta^{\ddagger}$	
Shear forces	From analysis	Consistent with section design for bending with $\gamma_{Rd} = 1{\cdot}20$	Consistent with section design for bending with $\gamma_{Rd} = 1{\cdot}35$
Critical region length	Max. of (d, $L/6$, 450 mm)	Max. of ($1{\cdot}5d$, $L/6$, 450 mm)	Max. of ($1{\cdot}5d$, $L/6$, 600 mm)
Min. mechanical volumetric ratio $\omega_{wd,min}$	0·05	0·09	0·13
Max. normalized axial force $v_{d,min}$	0·75	0·65	0·55
k_o (proportional to ω_{wd})	65	60	55
Design resistance verification	$0{\cdot}7\ M_{Rdi} > M_{Sd,CDi}$		Biaxial bending
Min. $\mu_{1/r}$ (proportional to ω_{wd})	5	9	13
Min. dimension	200 mm	250 mm	300 mm
Walls	**Ductility class L**	**Ductility class M**	**Ductility class H**
Critical height	Max. of (l_w, $H_w/6$) $< 2h_s$ and $2l_w$		
Design moments	'Tension shift' envelope		
Max. design shear force magnification factor ϵ	1·30	2·63	3·50
Min. $\mu_{1/r}$ (CCDF)	2·45	5·51	9·80
Min. web thickness	Min. of (150 mm, $ql_w/60$, $h_s/20$)		
Min. horizontal and vertical web reinforcement in critical height: %	0·2		
Min. total vertical reinforcement in critical height: %	0·4		
Boundary Elements	Within each ductility class, detailing requirements for longitudinal and transverse reinforcement as for columns with CCDF as above		
Min. boundary element length	$0{\cdot}15l_w$ or $1{\cdot}50b_w$		
Coupling beams	**Ductility class L**	**Ductility class M**	**Ductility class H**
Bi-diagonal reinforcement	Detailed according to column provisions if $V_S > 4bd\tau$		
Max. hoop spacing	100 mm		

Notes:

* 0·26% for C25/30 and S500 with $\rho'/\rho = 0{\cdot}5$ † 1·4% for C25/30 and S500 with $\rho'/\rho = 0{\cdot}5$

§ 0·8% for C25/30 and S500 with $\rho'/\rho = 0{\cdot}5$

** $\alpha_{CD} = \gamma_{Rd}\ (M_{Rdi,\ right\ beam} + M_{Rdi,\ left\ beam}) / |M_{Sdi,\ top\ column} - M_{Sdi,\ bottom\ column}|$

‡ $\delta = |M_{Sdi,\ right\ beam} - M_{Sdi,\ left\ beam}| / (M_{Rdi,\ right\ beam} + M_{Rdi,\ left\ beam})$

6. Assessment of existing buildings

6.1. Introduction

The deficiency in expected seismic performance of existing reinforced concrete buildings designed between 1930 and about 1975 (when design codes were implemented containing seismic provisions more or less equivalent to those currently in practice) has recently been recognized. In several European earthquake-prone countries, an unprecedented construction activity began after the second World War and lasted until the early 1980s. This has resulted in a major part of our building heritage being between twenty and forty years of age, and requiring renovation in most architectural and technological sub-systems, such as roofs, facades, window frames, plumbing, heating systems, etc. It is therefore believed that the building construction industry will be more active in renovation work than in new construction during the next decade, and this is confirmed by statistical observations on the market trends. This situation offers the unique possibility of retrofitting vulnerable buildings while other renovation work is taking place, with a very significant reduction in costs and difficulty. It is therefore necessary to define specific methods for assessing and strengthening reinforced concrete buildings designed between the 1950s and the 1970s, that can be identified as a class of earthquake risk buildings (ERBs) with common problems and deficiencies.

This chapter is focused on the evaluation of the structural capacity, or strength and deformability supply, rather than on demand, which obviously depends on the specific seismicity of the site. Also, the chapter is focused on assessment of the response, rather than on the evaluation of strengthening techniques.

With the category of ERBs considered in this chapter, deficiency of seismic performance is generally a consequence of lack of ductility rather than inadequate lateral strength. Seismic design coefficients in current codes generally imply dependable inelastic cyclic response to significant levels of ductility. In older buildings, the ductility deficit is a consequence of two major failings in the original design process — poor detailing of reinforcement and the lack of a capacity design philosophy. Deficiencies in detailing typically relate to amount, distribution, and anchorage of transverse reinforcement, although deficiencies in longitudinal reinforcement also exist. Frequently, transverse reinforcement in potential plastic hinge regions of beams, columns or walls is widely spaced, and anchored with 90° bends in the cover concrete. Spalling of compression concrete then leads to buckling of longitudinal reinforcement and collapse of the plastic hinge region. Shear reinforcement is also frequently inadequate, particularly in potential plastic hinge regions, where the strength of concrete shear-resisting mechanisms can be expected to reduce with increasing ductility, as inclined flexure–shear cracks increase in width, and aggregate interlock becomes increasingly ineffective. Beam–column joints were generally not designed with internal transverse reinforcement to carry the high shear stresses associated with moment reversal across the joint, resulting in a high potential for joint shear failure.

As a consequence of the lack of capacity design considerations in the design process, there is no assurance that a suitable hierarchy of strength exists to proscribe non-ductile modes of failure, such as shear failure, or limited ductile deformation mechanisms such as soft-storey sway mechanisms. Design to an allowable stress philosophy rather than a strength design philosophy, as was common before the late 1970s, contributes to the uncertainty of inelastic response.

Complete assessment of the expected seismic performance of these buildings is not simple. Current established procedures tend to be rather rudimentary, where details are compared with a check-list of possible deficiencies, and where calculations, if carried out at all, are of a simplistic nature, inadequate to determine the probable response (JBDPA, 1977, ATC21, 1988, ATC22, 1989, ATC14, 1982). On the other hand, the development of damage indices relating damage levels of specific classes of structures to seismic intensity based on experience in past earthquakes is being inappropriately used to determine seismic risk of individual buildings. It is clear that the application of a mean value from a data set with extremely wide scatter will provide little insight beyond indicating the need for more detailed structural calculations. Unfortunately, these risk analyses are being used routinely to guide retrofit decisions and strategies for specific buildings.

Traditional force-based seismic assessment would involve computation of member strengths and elastic period and a review of the detailing provided in the structure (Fig. 6.1). In the USA, depending on detailing and relative beam–column strengths, a decision would be made as to whether the frame should be considered an ordinary moment resisting frame (OMRF) or a special moment resisting frame (SMRF), and different force reduction factors, applied to the elastic 5% acceleration spectrum, determined accordingly (UBC, 1994).

In Europe the reference document would be the appropriate part of Eurocode 8 (EC8/1-4), which is focused on strengthening and repair rather than on assessment, and essentially refers to standard design methods for new buildings for analysis and verifications.

Clearly, the appropriate use of non-linear dynamic analysis is a viable method of approaching the problem of assessing the response of a building structure, with a series of difficulties related to the description of uncertain details, to the efficient mechanical modelling of inadequate design details and to the availability and user-friendliness of efficient computer programs.

A possible simplified approach to the global assessment of existing buildings can be based on a revisitation of the 'system approach' recommended by modern codes for the design of new buildings, where the flexural response assumed in design is forced by the application of appropriate capacity design (CD) principles. Clearly, in the case of existing buildings a desired response cannot be forced *a priori*, but a hierarchy of the members and mechanisms strengths can conveniently be used to assess the true deformation and energy dissipation capacity.

A first attempt to provide a meaningful system approach to the assessment of existing frame buildings was made by Priestley and Calvi (1991). A two-level seismic assessment procedure was outlined, intended to determine the risk, in terms of annual probability of exceedance, associated with both serviceability and ultimate limit states. Determination of the ultimate limit state involved an attempt to identify the most critical collapse mechanism, and calculation of its associated strength and ductility in system response terms. Strength and structural ductility were combined to provide an equivalent elastic response force level, which, by comparison with the design elastic response spectrum could be used to determine annual probability of exceedance corresponding to development of structural capacity. The basis for identifying the critical collapse mechanism was a modified form of capacity design, which permitted local element failure provided overall structural integrity was not jeopardized.

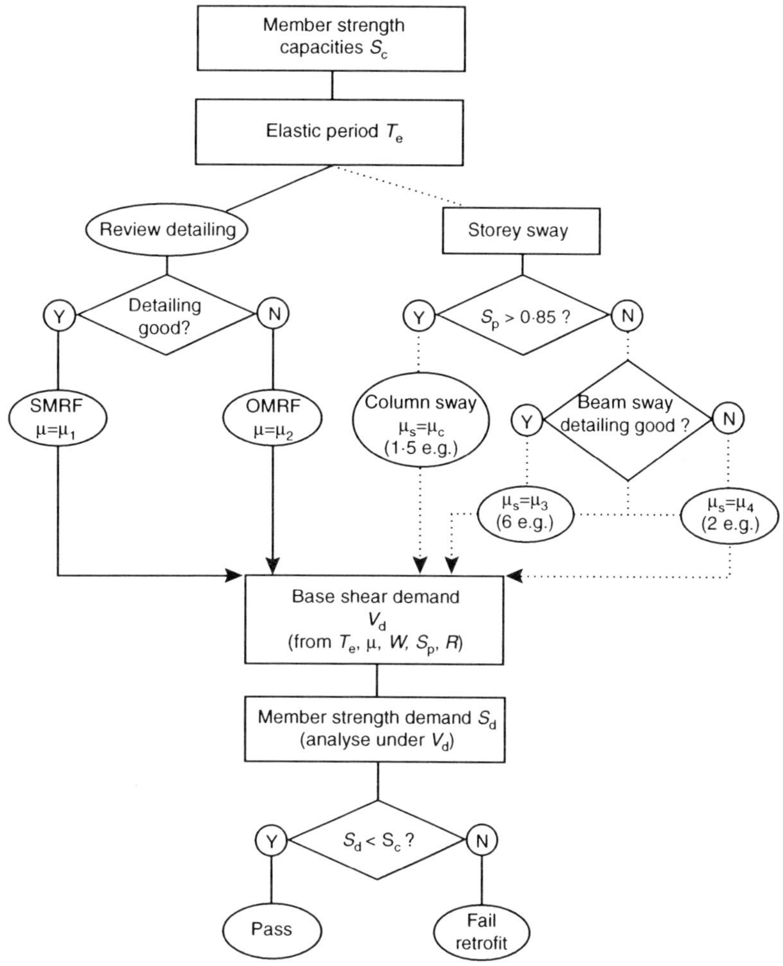

Fig. 6.1. Conceptual flow chart for a force-based assessment

The procedure has been re-examined by Priestley (1997), with systems response quantified directly in terms of structural displacement, instead of equivalent elastic strength. Strengths and deformation capacity of different critical mechanisms, in particular those relating to column shear strength and beam–column joint capacity were redefined, and ductility capacity of soft-storey mechanisms discussed in greater detail. The revisions to the earlier approach were, to a considerable extent, based on experience with seismic assessment of concrete bridges (Priestley and Seible, 1994, Priestley *et al.*, 1996). Suggestions relative to assessment of reinforced concrete buildings with structural wall bracing systems were also made.

In this chapter the concepts outlined above are revisited, with more emphasis on the different limit states applicable to existing structures and on which conditions correspond to each limit state. The role of capacity design principles in assessing the vulnerability of existing structures is also discussed in some detail.

The field of applicability of the concepts expressed is restricted to undamaged structures, while in the case of post-earthquake investigations added uncertainties have to be considered. On the contrary, procedures of the kind here described can conveniently be used for the preliminary verification of the design of new buildings. Finally, a warning has to be given regarding the effects of infills, not explicitly considered in this chapter, which can invalidate capacity design principles and induce less ductile failure modes in some cases, but in general tend to increase the seismic safety, when uniformly distributed throughout the structure.

Throughout this chapter European, American and New Zealand notations will be used as appropriate in relation to the formulations applied.

6.2. Limit states in assessment of existing structures

6.2.1. Knowledge and uncertainties in new design and assessment

From the point of view of an evaluation of the probability of attaining a given limit state, the conceptual difference between designing a new structure and assessing an existing structure consists of the different level of reliability of the knowledge related to some design aspects.

This is generally evident when considering that the design of new structures results essentially in a prescriptive definition of materials' properties, structure geometry and reinforcement details. The designer therefore has full control over the conceptual model of his product, but may have some doubt as to the correspondence between the model and the real structure. This is, of course, one of the main reasons for applying safety factors in the design process.

When the problem is to assess the probable response of an existing structure, the real object is available with, consequently, the possibility of checking the actual geometry and the quality of the construction process, as well as the material properties (which can be evaluated experimentally as briefly addressed in section 6.6). On the contrary, reinforcement details often have to be inferred, unless the original design plans are available.

The different level of knowledge for different design variables may result in different choices for the selection of models, different relative importance of limit states and CD principles and different protection factors against undesired events, particularly concerning CD principles.

Another important difference between new and existing structures is related to the potentially very different economical implications of modifying some design details and retrofitting an existing structure. These aspects are particularly relevant for serviceability and damage limit states, but substantially invalid when a collapse limit state is examined.

Finally, it is often mentioned that the remaining life of an existing structure is conceptually shorter than the design life of a corresponding new structure, and that it is therefore appropriate to modify the seismic action to take into account a reduced probability of a similar seismic event. This aspect is not relevant to an analysis focused on supply rather than on demand, but it is worth mentioning that the argument is more intellectual than real, due to the current erratic definition of design life for a building structure. A conceptual approach to considering a reduction of the design life for assessment is briefly presented in ERB (1996).

In the following sections, three limit states (LS) will be considered, essentially corresponding to collapse, reparability and possibility of immediate use after the seismic event. These limits do not necessarily correspond to the definition of current codes of practice. For example in EC8 two LS are considered, defined as 'ultimate' ('those associated with collapse or with other forms of structural failure which may endanger the safety of people') and 'serviceability' ('those associated with damage occurrence, corresponding to states beyond which specified service requirements are no longer met').

6.2.2. Ultimate limit state

It is often said that collapse of a structure should not take place during the strongest ground shaking considered feasible for the site. Protection against loss of life is the prime concern here and equally high priority must be accorded in designing new structures and assessing the response of existing structures. Extensive damage may have to be accepted under an earthquake corresponding to the ultimate limit state, to the extent that it may not be

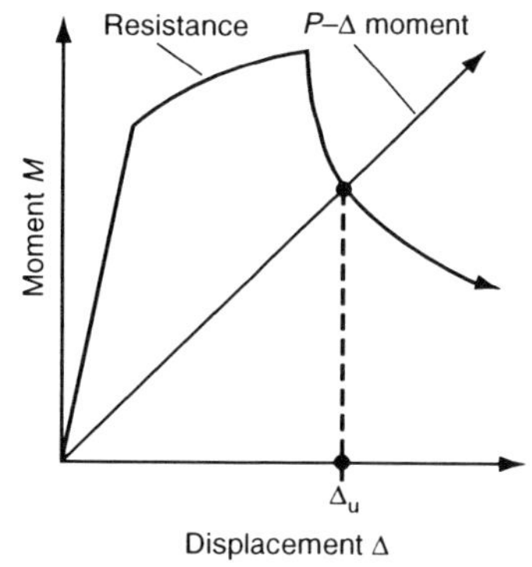

Fig. 6.2. P–Δ collapse of a structure

economically or technically feasible to repair the structure after the earthquake. Demolition and replacement may be required.

The determination of the situation corresponding to this LS is of critical concern, but is also very difficult, requiring the modelling of structural response up to a real collapse. It is actually clear from an in-depth analysis that the clauses of most codes of practice for ultimate LS correspond to conventional definitions of collapse, more appropriately related to some heavy damage LS, and are still more oriented to each single structural member than to an overall response.

In the case of existing structures the assessment of a survival LS is of fundamental importance, at least from a philosophical point of view, since it is the only LS for which a trade-off between the cost of a preliminary retrofitting and global cost of repair, replacement and interruption in the use of the structure cannot be accepted.

It is almost a truism to state that this LS corresponds to the condition when the structure is no longer able to support gravity loads and therefore collapses, but this is nonetheless a very valuable and effective way of defining the LS. Even when the lateral resistance of a critical section has been substantially reduced, the structure may still be stable. Collapse of a structure will occur when gravity-load capacity is reduced below the level of existing gravity loads as a result of, say, total shear failure or disintegration of a column plastic hinge. Alternatively, collapse can result from a stability failure, where P–Δ moments exceed the residual capacity of the columns. As shown in Fig. 6.2, if the ultimate displacement capacity assessed from the intersection of the resistance and P–Δ curves exceeds the maximum expected displacement in the survival-level earthquake, collapse should not occur.

6.2.3. *Reparability limit state*

The potentially severe economical implications of an intervention on an existing structure may raise the importance of a damage limit state where a certain amount of repairable damage is permissible. For a new structure, a damage LS implies a design-level earthquake of reduced probability of occurrence compared to the serviceability LS; in the case of existing structures this is not a necessary condition, depending on comparison between cost of retrofitting and global (including interruption of use) cost of repair.

The permitted damage may include spalling of cover concrete requiring cover replacement, and the formation of wide flexural cracks requiring injection grouting to avoid later corrosion problems. However, the essential aspect of response to this limit state is that the required repair should be repairable. Fracture of transverse reinforcement or buckling of longitudinal reinforcement should not occur, and the core concrete in plastic hinge regions should not need replacement. With well-designed structures, this limit state generally corresponds to displacement ductility factors in the range of 3–4.

In some cases, beyond this LS the lateral resistance may start to diminish with increasing displacement, collapse being not far off.

This LS is probably the most important in terms of seismic assessment and most of the quantitative assessment indications given in this chapter are related to it.

6.2.4. *Immediate-use limit state*

A structural situation for which the structure will be immediately usable after a seismic event, without any need of repair, is often defined as a serviceability LS. Member flexural strengths could be reached, and some limited ductility developed, provided that concrete spalling in plastic hinges did not occur and that residual crack widths remain sufficiently small so that remedial activity, perhaps in the form of epoxy injection of cracks, is not needed. For assessment, limit values of $\epsilon_c = 0{\cdot}0035$–$0{\cdot}004$ and $\epsilon_s = 0{\cdot}01$–$0{\cdot}015$ are suggested. Note that similar values are normally considered appropriate for an 'ultimate limit state' when designing for gravity loads. The validity of these strain limits can be determined as follows. Typically, spalling of concrete is initiated at extreme fibre compression strains between $\epsilon_c = 0{\cdot}006$ and 0·01 (Mander *et al.*, 1988). Thus, the limit of 0·004 is a conservative estimate of the onset of structural damage. The strain limit of $\epsilon_s = 0{\cdot}015$ has been determined to ensure that residual crack widths do not exceed 1·0 mm. Considering a typical plastic hinge region in a column where cracks form at an average spacing of 200 mm, the crack width corresponding to an extreme reinforcement strain of 0·015 will be $0{\cdot}015 \times 200 = 3{\cdot}0$ mm. However, this crack width is the value at maximum response. For low ductility levels, residual displacements (and hence crack width) are approximately one-third of maxima values, and hence a residual crack width of 1·0 mm can be expected. This is frequently taken as the maximum width that can be tolerated in normal environmental conditions without requiring remedial action.

6.3. Methods of assessment

In this section the relative merits of force- and displacement-based approaches for the assessment of existing structures will be discussed. In both cases it is assumed that it is possible to define an equivalent single-degree-of-freedom structure, characterized by an appropriate force–displacement response.

6.3.1. *Force-based system assessment*

A traditional, force-based, assessment procedure is commonly based on determining the probable strength of the critical collapse mechanism, the secant stiffness to a conventional yield point, and the available global displacement ductility. It is therefore necessary to estimate the probable flexural and shear strengths of sections, members and joints and to determine the resulting post-elastic global deformation mechanism. The fundamental period of vibration of the structure will be calculated considering the secant stiffness to the conventional yield point.

The definition of period of vibration, available lateral strength and structural displacement ductility, allow the assessment of the likely seismic performance of the structure entering the appropriate class of acceleration spectra, characterized by the local seismicity, and by the required structural ductility factor.

An assessment procedure can be summarized in the following steps, which correspond to the process shown in Fig. 6.1.

(*a*) Estimate the probable flexural and shear strength of the critical sections of members and joints.
(*b*) Determine the probable post-elastic mechanism of deformation of the structure, the corresponding strength V, and the corresponding global displacement ductility capacity μ, as a function of the limit state under consideration (e.g. a different ductility will be permitted when assessing a collapse and a reparability limit state).
(*c*) Determine the conventional elastic period of vibration of the structure T.
(*d*) Estimate the acceleration coefficient that the structure can tolerate, dividing the strength by the seismic weight of the structure, corrected by the appropriate factors (importance, occupancy and similar).
(*e*) Enter the local acceleration spectra, characterized by different levels of required ductility and by the appropriate return period (depending on the limit state under consideration).
(*f*) If the point corresponding to the structure period and acceleration coefficient falls below the spectrum whose required ductility corresponds to the structure available ductility, the structure needs to be strengthened, either increasing its ductility capacity or its strength (in both cases with the possible implication of changing its post-elastic deformation mechanism).
(*g*) If the structure is safe from the point of view of strength and ductility capacity, estimate the inter-storey drift and verify if it is acceptable in terms of the requirement of the limit state under consideration.

A procedure based on this kind of approach was discussed by Priestley and Calvi (1991), who proposed a method to determine an equivalent elastic acceleration for the structure, corresponding to ultimate capacity of the critical inelastic deformation mechanism, from the lateral strength and the structure displacement ductility capacity. The method for relating this to the equivalent elastic acceleration response $S_{a(e)}$ assumed that the latter could be found from the relationship

$$S_{a(e)} = R \times S_{a(mech)} \tag{6.1}$$

where

$$R = 1 + (\mu_s - 1)T/1{\cdot}5T_0 \leq \mu_s \tag{6.2}$$

and where $S_{a(mech)}$ is the acceleration coefficient for the structure corresponding to the mechanism strength, μ_s is the structural displacement ductility capacity corresponding to the mechanism investigated, T is the elastic period of the structure, and T_0 is the period corresponding to peak spectral response (see Fig. 6.3(a)). Equations (6.1) and (6.2) imply that the 'equal displacement' approximation (i.e. $R = \mu_s$) of structural response applies for $T \geq 1{\cdot}5T_0$, and that the response changes linearly from the 'equal acceleration' approximation ($R = 1$) at $T = 0$, through the 'equal energy' approximation ($R = \sqrt{(2\mu_s - 1)}$) at about $T = 0{\cdot}7T_0$, to the equal displacement approximation at $T \geq 1{\cdot}5T_0$.

The weakness of the above approach is the assumption concerning the relationship between ductile and equivalent elastic response (equal-energy, equal-displacement, etc.), and the lack of consideration of hysteretic energy dissipation characteristics. Conceptually, it also places undue emphasis on strength, which is considered as the main parameter to be corrected using the appropriate behaviour factor. Since in ductile systems failure occurs not when the strength is reached, but when the ductility capacity (i.e. the ultimate displacement) is reached, it may make more sense to compare demand and capacity directly in terms of displacements. When the response

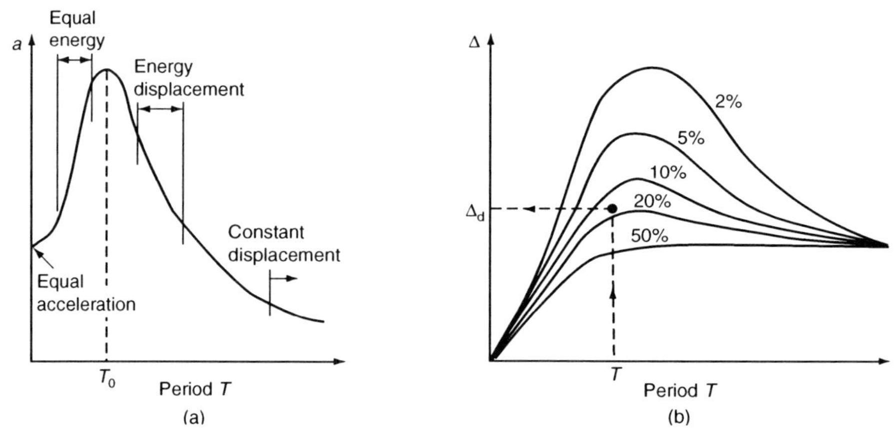

Fig. 6.3. Qualitative design spectra for assessment: (a) acceleration spectra; (b) displacement spectra

is purely elastic, it is of equal validity to compare response in terms of displacement rather than acceleration.

6.3.2. Displacement-based system assessment

Response can be considered directly in terms of displacement, using the substitute-structure approach of Shibata and Sozen (1976). In this, the structural period T is not related to the initial elastic stiffness k_i but to the effective stiffness k_{eff} at maximum displacement, as shown in Fig. 6.4(a). Thus

$$T = 2\pi\sqrt{\frac{M}{k_{eff}}} \tag{6.3}$$

Maximum displacement demand corresponding to the design or assessment requirement, is found from a set of displacement response spectra, for different levels of equivalent viscous damping as shown, for example, in Fig. 6.4(b). The level of damping assumed depends on the structural displacement ductility demand μ_s, and the predominant form of plastic hinging developed. As shown in Fig. 6.4(b), the energy dissipated in beam plastic hinges is typically larger than in column plastic hinges, and this should be recognized in the estimation of equivalent viscous damping.

Thus, seismic response is characterized by an equivalent elastic stiffness and damping corresponding to maximum response, rather than initial values, based on k_i and 5% damping, as typically used in force-based design or assessment.

With equivalent period and damping calculated, the required displacement Δ_d is read from the displacement spectra (Fig. 6.3(b)), and

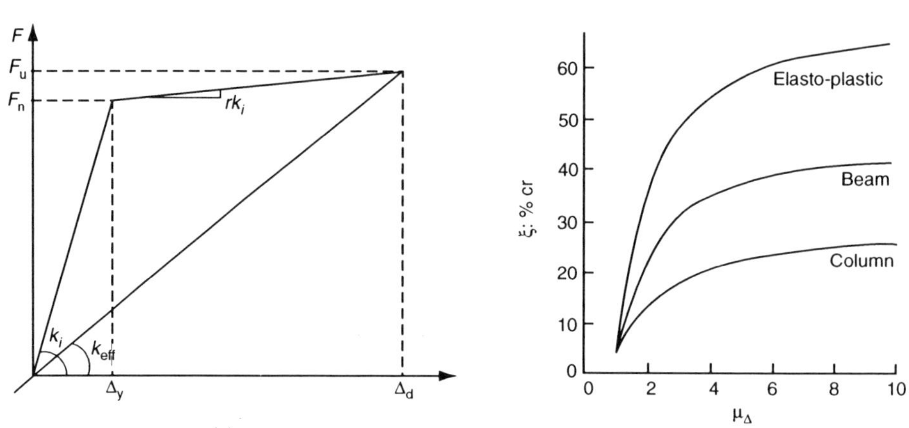

Fig. 6.4. Elements for the definition of an equivalent model in displacement-based assessment: (a) effective stiffness; (b) equivalent viscous damping

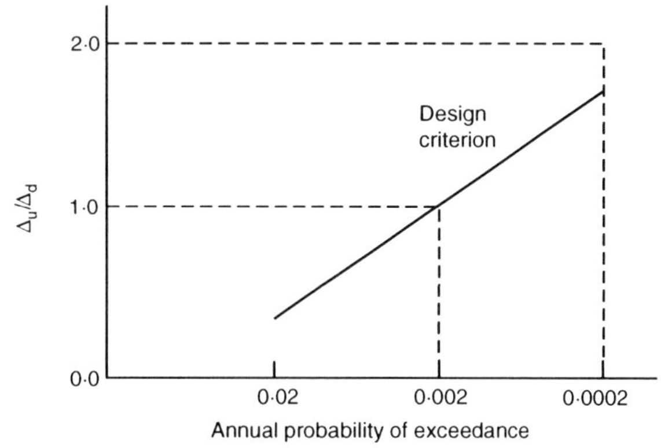

Fig. 6.5. Relationship between annual probability of exceedance and displacement ratio

compared with the ultimate displacement capacity Δ_u. Seismic risk associated with achieving the ultimate displacement of the structure is then assessed from a relationship between the ratio Δ_u/Δ_d and annual probability of exceedance, shown conceptually in Fig. 6.5, where a value of $\Delta_u/\Delta_d = 1$ corresponds to an annual probability of exceedance of 0·002, or a return period of about 500 years — a common value used for design of new structures. The acceptable minimum level for Δ_u/Δ_d is thus related to seismic risk.

The approach outlined above has the advantage of directly considering displacements, which can be related to strain-based limit states. These are clearly more fundamental to damage than force-based limit states, even when serviceability, rather than ultimate conditions are considered. For example, it is currently common to define the serviceability limit state as corresponding to the limit of 'elastic' response, or to a structure displacement ductility demand of $\mu_s = 1$ (these are not necessarily the same). Actually, the onset of damage at a level requiring repair, and thus influencing serviceability, as evidenced by unacceptably wide crack widths and/or concrete spalling, will rarely correspond to such a low level of response, depending critically on such aspects as axial load level in members, and type of inelastic deformation mechanism achieved. A serviceability limit corresponding to $\mu_s = 1$, while generally (although not exclusively) conservative, provides a very uneven level of protection against damage. Use of a displacement-based assessment procedure, where the structural displacement corresponding to development of serviceability limit strains (say peak concrete compression strain = 0·004, or peak reinforcement tensile strain = 0·01), enables a consistent level of assessment to be achieved.

The above discussion indicates that, for the case of a serviceability LS, the conceptual difference between the two methods results practically in a different evaluation of the acceptable displacement, and therefore of the equivalent secant stiffness.

It also has to be noted that direct calculation of displacement or drift limits reduces the potential compounding of uncertainties and errors, which are amplified by the multiplication of elastic displacement and ductility values.

The procedure recommended above requires generation of assessment displacement spectra representing acceptable performance. In theory, these can simply be generated from acceleration spectra using the normal relationships between peak acceleration and displacement of elastic oscillators. However, it should be recognized that design acceleration spectra are often unrealistically high in the long period range. This is typically a consequence of a deliberate decision to enforce minimum

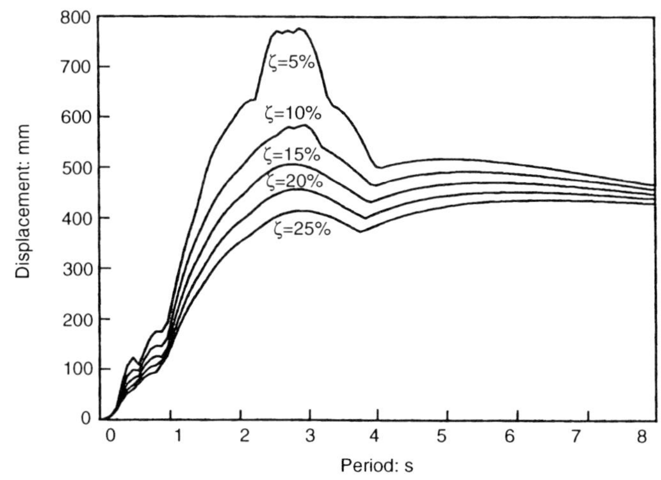

Fig. 6.6. Displacement spectra for 1994 Northridge earthquake (Sylmar record)

strength requirements for long period structures, rather than a reflection of expected seismic response. Determination of spectral displacements from design acceleration spectra therefore typically results in displacements which continue to increase with period even at very large values of T, although it is known that the characteristics shown in Fig. 6.3(b) are more realistic. Here, spectral displacements reach a maximum, and then decrease again at large periods, eventually reaching a stable value equal to the peak ground displacement, regardless of ductility level or period. The implication is that the equal displacement approximation can be expected to be excessively conservative in the long period range. The influence of this conservatism in existing spectra will be greater in a substitute structure analysis, based on effective period at maximum response, than in an initial-stiffness force-based analysis. Consequently, for seismic assessment, it is important to use a set of displacement spectra that is as realistic as possible.

The characteristics of 'real' displacement response spectra are shown, for example, in the Sylmar record from the 1994 Northridge earthquake, recorded close to the epicentre, which displayed a peak ground acceleration of 0·85 g. As shown in Fig. 6.6, a peak spectral displacement of about 750 mm occurred at about $T = 2{\cdot}5$ s, for 5% damping. However, for equivalent viscous damping of about 20–30%, which is typical of stable hysteretic response of systems with beam and column hinges, the response is essentially flat at about 450 mm for $T \geq 2$ s. The implication is that satisfactory response of longer period structures could be assured under this extreme record, provided a displacement capacity of $\Delta_u > 450$ mm was available, regardless of available strength.

For regions of only low to moderate seismicity, such as much of Europe, this approach may have increased significance in assessment, since the M5 to M6 earthquakes, which could be considered to represent extreme events in many cases, may have high peak spectral response accelerations, but rather low peak spectral displacements. It would therefore seem that flexible structures reasonably designed for gravity loads are unlikely to be at significant risk of collapse in such cases.

The assessment procedure thus follows the sequence of operations in the flow chart of Fig. 6.7. Calculated member strengths are used to determine the storey sway potential S_p, discussed subsequently, and hence, the inelastic deformation mechanism and base shear strength V_b. Member plastic rotation capacities are calculated from moment–curvature analyses. Shear strength of members and joints is checked to determine whether shear failure will occur before the limits to flexural plastic rotation are obtained, and the available plastic rotation capacity is reduced, if necessary, to the

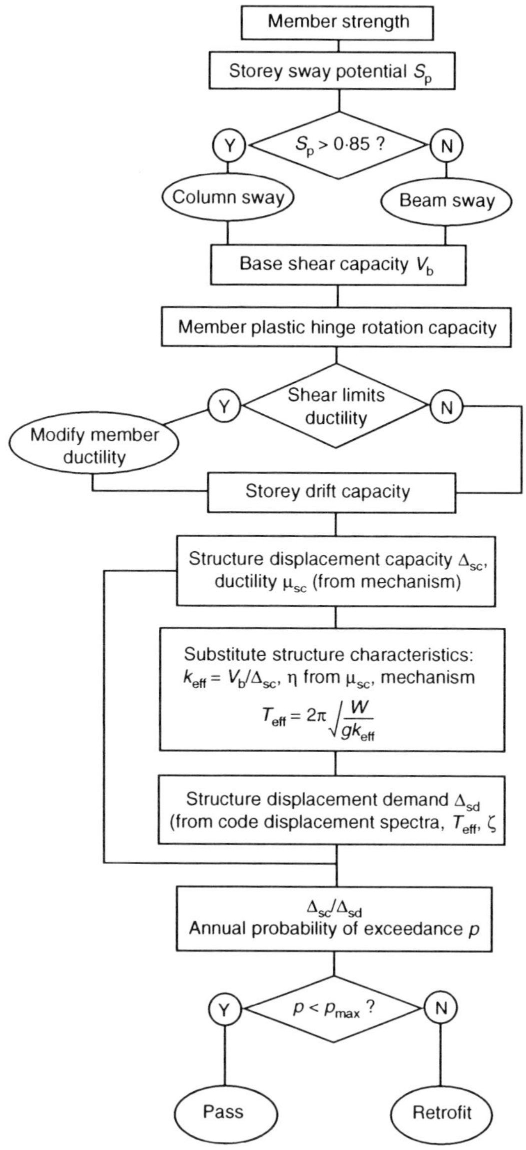

Fig. 6.7. Conceptual flow chart for a displacement-based assessment

value pertaining at shear failure. From the plastic rotation capacities of members in a storey, the plastic drift is estimated. The structure displacement capacity Δ_{sc}, and ductility capacity, μ_{sc}, are found from the mechanism of plastic deformation determined early in the analysis, and the critical storey drift.

The effective stiffness at maximum displacement, $k_{eff} = V_b/\Delta_{sc}$, and corresponding period of vibration $T_{eff} = 2\sqrt{[W/(gk_{eff})]}$ are calculated. The equivalent viscous damping ζ, is based on the structure ductility, and mode of inelastic displacement. The structure displacement demand Δ_{sd} is then found, entering the code displacement spectra with T_{eff} and ζ. Comparison of displacement capacity and demand enables the risk, in terms of annual probability of exceedance, to be identified. The procedure is linear, and needs no iteration.

In analogy with the force-based case, and with reference to Fig. 6.7, the displacement-based assessment procedure discussed in this section can be summarized in the following steps.

(*a*) Estimate the probable flexural and shear strength of the critical sections of members and joints.

(*b*) Calculate the member plastic rotation capacity from moment–curvature analysis, considering the appropriate values for the limit state under consideration. Reduce the plastic rotation capacity of the members where it is limited by the shear strength of members and joints.
(*c*) Estimate the storey plastic drift capacity from the plastic rotation capacities.
(*d*) Determine the probable post-elastic mechanism of deformation of the structure, the corresponding strength V, the overall structure displacement capacity Δ and the corresponding global displacement ductility capacity μ.
(*e*) Calculate the effective stiffness at maximum displacement, and the corresponding effective period of vibration of the structure T.
(*f*) Determine the equivalent viscous damping, considering the estimated ductility capacity and the expected mechanism (the cyclic dissipation capacity of columns is lower than that of beams).
(*g*) Estimate the structure displacement demand entering the appropriate displacement spectrum, characterized by different levels of equivalent viscous damping and by the appropriate return period (depending on the limit state under consideration).
(*h*) Compare the displacement capacity and the displacement demand.

6.3.3. Comparison between force-based and displacement-based assessment

For the medium to long period range of response, force-based procedures typically assume the validity of the 'equal displacement' approximation that maximum displacement of an inelastic structure is identical to that of an elastic structure with the same initial period, regardless of the level of ductility and energy absorption characteristic of the inelastic system. In the displacement-based procedure, displacement demand is based on equivalent structural stiffness at maximum displacement and an estimate of the equivalent viscous damping.

In order to emphasize differences (and similarities, where appropriate) between the two approaches, displacement spectra compatible with the New Zealand Loadings Code (1992) have been generated for the three soil conditions identified in the code, and are plotted in Fig. 6.8. These have been generated as follows. The 5% displacement spectra for $\mu = 1$ have been obtained from the NZS 4203:1992 acceleration spectra as

$$\Delta_{(T)} = 9800C_{(T,1)}T^2/4\pi^2 \quad \text{(mm)} \tag{6.4}$$

for the three soil categories, where $C_{(T,1)}$ is the period-dependent acceleration coefficient for displacement ductility $\mu = 1$.

Displacement spectra for damping values different from 5% have been obtained by multiplying the displacement given by equation (6.4) by the factor

$$K_\zeta = \left(\frac{7}{2+\zeta}\right)^{\frac{1}{2}} \tag{6.5}$$

Equation (6.5) has been used in developing displacement spectra for the European Seismic Code EC8 (1994).

Examination of the displacement spectra in Fig. 6.8 reveals that, apart from some non-linearity for low periods, particularly for flexible soils, the curves are well represented by straight lines from the origin. This is illustrated by dashed straight lines for intermediate soils in Fig. 6.8(b). It should be noted that if the 5% damping curves had been generated as

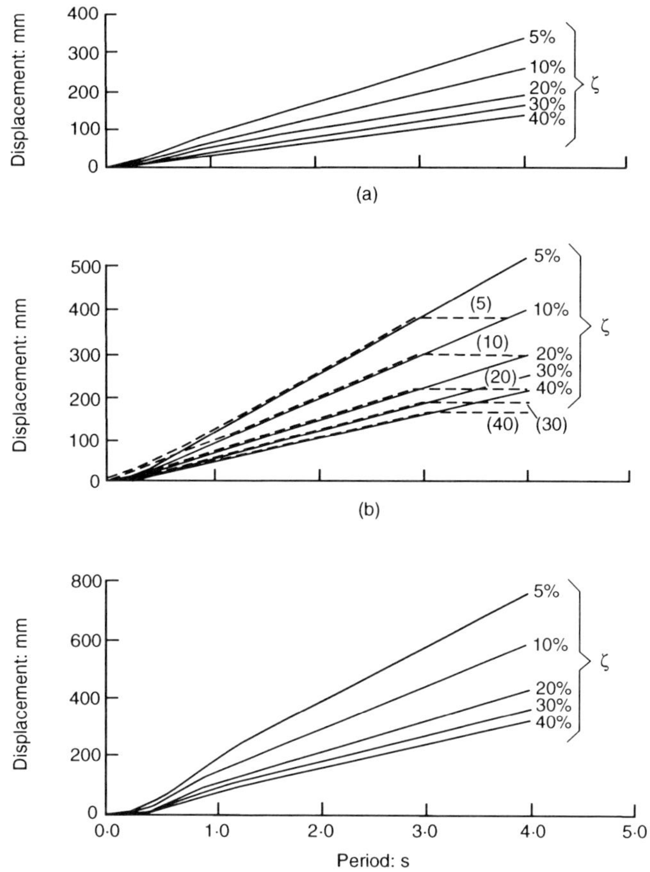

Fig. 6.8. Displacement spectra to NZS 4203 : 1992 and equation 6.4: (a) rock and firm soil; (b) intermediate soils; (c) flexible or deep soil sites

$$\Delta_{(T)} = \mu 9800(C_{T,\mu})T^2/4\pi^2 \tag{6.6}$$

which, using the force-based approach would be appropriate for ductile response, the deviation from the straight lines in the low period range would be further reduced. Since only structures with initial period of vibration longer than 0·7 s will be considered in this section, it would seem quite adequate to represent the displacement spectra by straight lines.

It is also apparent that the acceleration spectra imply displacements that keep increasing linearly as the period increases. Examination of recorded spectra indicate this to be unrealistic (see Fig. 6.5 for example). This is recognized in EC8 which applies a displacement cut-off at $T = 3$ s, as suggested by the linked dash-dot lines in Fig. 6.8(b). Note that this still does not represent the tendency of the spectra to converge at the peak ground displacement at high periods, but conservatively maintains the spectra at constant peak displacement response values for periods higher than those at peak displacement response. The approach should therefore be conservative.

Figure 6.8 enables comparisons between displacements predicted by force-based and displacement-based assessment to be made, using typical code spectra. With respect to Fig. 6.9, the required structure displacement for a building of elastic period T_e (and corresponding elastic stiffness k_e) is given by the 5% displacement spectrum of either Fig. 6.8(a), (b) or (c), depending on the soil type, independent of the ductility demand or energy dissipated, provided $T > 0{\cdot}7$ s. For the displacement-based procedure, with the same base shear strength and a displacement ductility factor of μ, the effective stiffness at maximum response (see Fig. 6.9) is

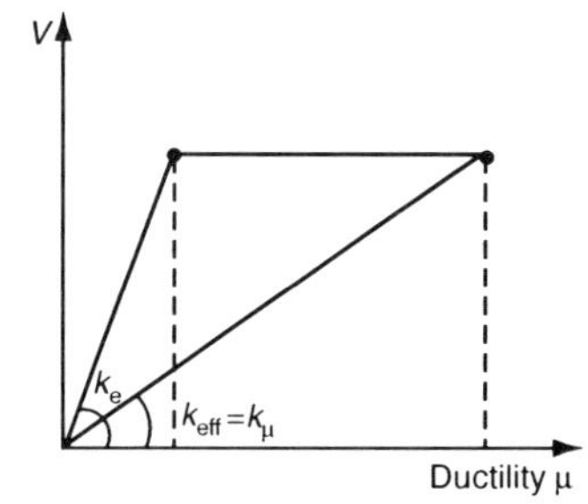

Fig. 6.9. Stiffnesses for force- and displacement-based assessment

$$k_{\text{eff}} = k_\mu = \frac{1}{\mu} k_\text{e} \tag{6.7}$$

The effective period at peak response T_{eff} is therefore

$$T_{\text{eff}} = \sqrt{\mu} T_\text{e} \tag{6.8}$$

The expected displacement is thus found for the specified soil condition from Fig. 6.8, using T_{eff}, and the damping appropriate for the ductility μ, and mechanism of inelastic response.

Table 6.1 shows results of a comparison of force- and displacement-based assessment of required structure displacements for initial elastic periods between 0·7 and 2·0 s, for the three soil conditions of Fig. 6.8. A single value of displacement is predicted from the force-based approach for each given period. Four different values are listed for the displacement-based approach. The first, for ζ = 10%, is appropriate for a non-linear elastic system such as that based on unbonded prestressing (Priestley and McRae, 1996). The second, for ζ = 20%, is appropriate for a structural wall building with μ_s = 4. The third and fourth, with ζ = 30% and 40%, are approximately appropriate for frame buildings with beam hinge mechanisms, and structural ductility factors μ_s = 3 and μ_s = 6, respectively (see, Fig. 6.3(b)).

Examination of Table 6.1 indicates that, in comparison with the displacement-based approach, the force-based prediction underestimates the displacement demand of the non-linear elastic and structural wall buildings (by 10–40%), overpredicts the frame building response at μ_s = 3 (by 10–20%) but gives a close estimate of the displacement for the frame building with μ = 6. Further, if the 3 s cut-off on displacement adopted in EC8 (1994) is considered, the force-based approach grossly overestimates the displacement demand for the T_e = 2 s, and somewhat overestimates the displacement demand for the T_e = 1·5 s frame structures.

In the absence of results obtained from reliable non-linear analyses or experimental tests, these comparisons do not necessarily indicate whether the displacements predicted by either one of the methods are closer to reality than those predicted by the other one. Some results available in the literature however (Calvi and Kingsley, 1995, Priestley *et al.*, 1996), seem to indicate a superiority of the displacement-based approach.

It is also emphasized that this comparison is based within the framework of the NZS4203 (1992) spectra, and the adjustment provided by equation (6.5).

Since the spectra are not felt to be totally appropriate in shape, the conclusions drawn above should only be taken as indicating trends. However, it is apparent that the displacement-based approach results in a range of possible displacements, depending on mechanism of inelastic response, which is not available in the force-based approach.

Table 6.1. Structure displacement demand (mm) to NZ Loading Code (NZS4203, 1992), according to force-based or displacement-based approaches

Spectra	Elastic period:	Force-based	Non-linear elastic $\mu = 3$, $\zeta = 10\%$		Structural wall $\mu = 4$, $\zeta = 20\%$		Beam hinge mechanism $\mu = 3$, $\zeta = 30\%$		Beam hinge mechanism $\mu = 6$, $\zeta = 40\%$	
	s	Δ_f	Δ_d	Δ_d/Δ_f	Δ_d	Δ_d/Δ_f	Δ_d	Δ_d/Δ_f	Δ_d	Δ_d/Δ_f
Rock	0·7	58	77	1·33	67	1·15	48	0·82	60	1·03
	1·0	82	112	1·37	95	1·15	69	0·84	81	0·99
	1·5	129	193	1·51	138	1·07	99	0·77	124 (101)*	0·96 (0·78)*
	2·0	169	216 (187)*	1·28 (1·11)*	185 (138)*	1·09 (0·82)*	134 (116)*	0·79 (0·68)*	166 (101)*	0·98 (o·60)*
Intermediate soil	0·7	79	115	1·46	96	1·22	68	0·86	85	1·08
	1·0	124	160	1·29	139	1·12	99	0·80	120	0·97
	1·5	184	243	1·32	213	1·16	153	0·83	190 (156)*	1·03 (0·85)*
	2·0	248	340 (289)*	1·37 (1·17)*	289 (213)*	1·17 (0·86)*	207 (178)*	0·83 (0·72)*	260 (156)*	1·05(0·63)
Soft soil	0·7	115	175	1·53	151	1·32	108	0·95	134	1·17
	1·0	186	251	1·35	211	1·14	155	0·83	187	1·01
	1·5	290	368	1·27	313	1·08	228	0·78	283 (229)*	0·97 (0·79)*
	2·0	377	493 (425)*	1·30 (1·13)*	423 (313)*	1·12 (0·83)*	305 (263)*	0·81 (0·70)*	379 (229)*	1·01 (0·61)*

* Numbers in parentheses based on displacement cut-off of $T = 3$ s.

6.4. Mechanism considerations

A fundamental aspect of seismic assessment is the identification of the probable inelastic deformation mechanism. This requires comparison of flexural and shear strength of members, to determine whether flexural or shear failure is anticipated, and comparison of the relationship between strengths of beams and columns framing into joints, to determine whether beam-sway or column-sway mechanisms are likely to form. The problem of sway mechanism determination is discussed first.

In a full capacity design procedure, column moment capacities are required to have a substantial margin of strength over beam capacities framing into the same joint, to ensure that the desired weak beam–strong column performance develops, thus proscribing column-sway mechanisms. If column hinges are to be completely avoided, then the margin of strength must reflect the influence of higher mode response as well as potential beam flexural overstrength (Paulay and Priestley, 1992). However, less conservative measures are appropriate if individual column hinging is permitted, provided that a full storey column-sway mechanism does not develop. A column-sway mechanism involves the formation of plastic hinges at the top and bottom of all columns at one level of a frame, as shown, for example, in Fig. 6.10(b). Thus, formation of individual column hinges should not be seen as particularly serious, since some flexural ductility capacity of the column hinges will exist, and only very minor plastic rotations can develop in the column hinges until a full storey mechanism develops.

As discussed by Priestley and Calvi (1991), the potential for developing a column-sway mechanism can be determined from the value of a sway potential index S_p. This is defined by comparing the flexural capacities of beams and columns at all joints at a given level n of the frame:

$$S_p = \frac{\sum_{i=1}^{j}(\Sigma M_{(Bn,i)})}{\sum_{i=1}^{j}(\Sigma M_{(Cn,i)})} \qquad (6.9)$$

where $\Sigma M_{Bn,i}$ = sum of beam moment capacities (left + right) at the joint centroid of joint i, level n, and $\Sigma M_{Cn,i}$ = sum of column moment capacities (upper and lower) at the joint centroid of joint i, level n.

If $S_p > 1$, a high probability exists of a column-sway mode forming, particularly if $S_p > 1$ at the first-floor level, since hinging must be expected at the base of the columns at ground-floor level. However, since the consequence of a soft-storey mode is a greatly reduced structural plastic displacement capacity, as compared to a beam-sway mode as discussed subsequently, a change from $S_p = 0{\cdot}98$ to $S_p = 1{\cdot}02$ would imply great variation in capacity. To avoid uncertainties in material properties and small errors in calculations unduly influencing the predicted sway mode, and to provide some recognition of higher mode effects, it is recommended that a column-sway mechanism be assumed to develop when $S_p > 0{\cdot}85$. Where

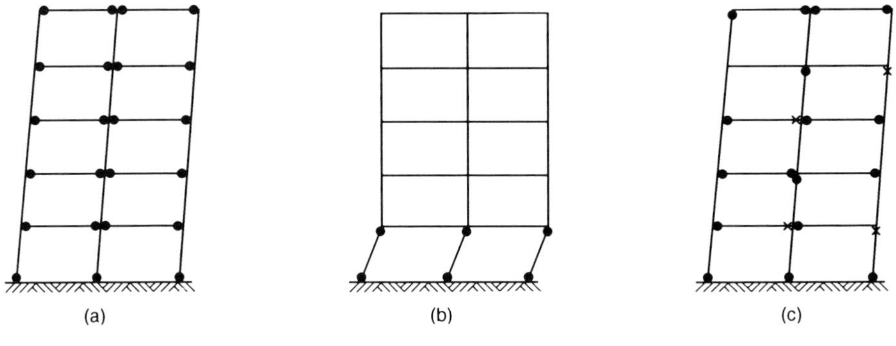

Fig. 6.10. Plastic collapse mechanisms: (a) beam sway; (b) column sway (soft storey); (c) mixed mode

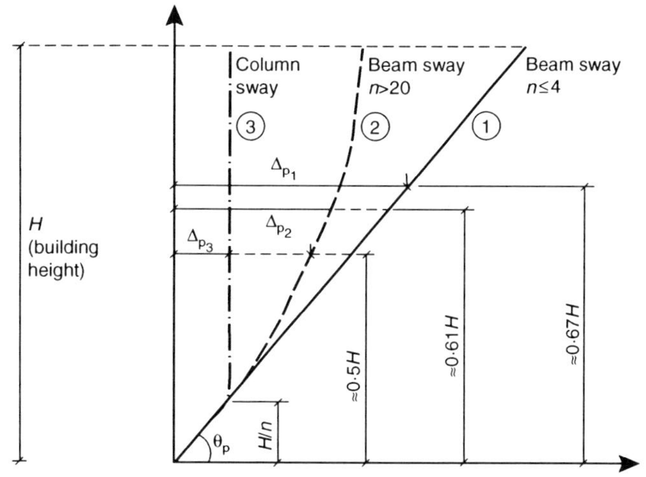

Fig. 6.11. Plastic displacement profiles

considerable vertical irregularity of the structural system exists, a lower value for S_p may be appropriate.

Having determined the expected sway mechanism, and its inelastic deformation capacity (as subsequently discussed) in terms of storey drift, the structural displacement at the centre of seismic force may be found. This requires investigation of the shape of the plastic deformation profile, as shown in Fig. 6.11. Ideally, this should be found from an inelastic frame lateral response analysis, incorporating all potential member non-linearities. This can be achieved using special purpose 'push analysis' programs, or by use of dynamic inelastic time-history analyses, where the lateral force vector is gradually increased in magnitude sufficiently slowly to ensure that dynamic modes of the structure are not excited. However, this assumes a knowledge of the shape of the lateral force vector, which will typically be assumed to be an inverted triangle, and which may be a reasonable approximation of the elastic displacement profile. If an inelastic deformation mode develops with a displaced shape markedly different from the assumed inverted triangular shape, as would be the case for a soft-storey column-sway mode, the vertical distribution of forces in the lateral force vector would gradually deviate (increasingly) from the inverted triangle shape. To warrant the sophistication of an inelastic static mechanism, or push analysis, it would seem that it would be necessary to be able to modify the shape of the lateral force vector, as plastic displacements increase. Although not conceptually difficult to implement in a push analysis, it is not thought to be currently available in any computer program.

The considerations discussed above are, however, relatively straightforward to implement in a hand analysis, although the degree of precision must be recognized as being rather coarse. Since our ability to determine realistic characteristics for design (or assessment) seismicity is of considerably greater coarseness, this should not be seen to invalidate this simple process.

Consider the plastic displacement profiles of Fig. 6.11. Three cases are considered, all with the same maximum plastic drift angle θ_p, assumed to develop in the lowest storey. The linear profile 1 corresponds to a beam-sway mechanism in a low-rise frame (say $n \leq 4$). For much taller frames (say $n \leq 20$) dynamic inelastic analyses indicate that at peak response, the plastic displacement profile is non-linear, with larger plastic drifts occurring in the lower floors. Paulay and Priestley (1992) recommend a peak plastic drift equal to about twice the average over the building height, although there is some evidence that this may be excessive when hysteretic characteristics are used that are more representative of reinforced concrete behaviour than the elasto-plastic analyses used as a basis for those recommendations. Profile 2 shows the expected shape, assumed to be parabolic. If a column-sway mechanism develops in the lowest floor, the plastic displacement shape is represented by profile 3.

Based on these shapes, the plastic displacement of the centre of seismic force can be estimated. First, however, it must be recognized that the centre of seismic force itself depends on the displaced shape. If an inverted triangle shape is a reasonable approximation of the elastic displacement response, then, initially the effective height of the single degree of freedom representation of the structure is approximately

$$h_{\mathrm{eff}} = 0{\cdot}67H \qquad (6.10)$$

This is also the effective height for the plastic displacement profile 1 of the short frame, but profiles 2 and 3 have shapes with lower centroids ($h_{\mathrm{eff}} \approx 0{\cdot}61H$, $h_{\mathrm{eff}} \approx 0{\cdot}5H$, respectively). Thus, as inelastic displacement increases, the centre of seismic force gradually decreases from $h_{\mathrm{eff}} \approx 0{\cdot}67H$ at $\mu_{\mathrm{s}} = 1$ to $h_{\mathrm{eff}} \approx 0{\cdot}61H$ (or $0{\cdot}5H$) at very large values of μ_{s}.

For the beam-sway mechanisms, the effect is not particularly significant, and it is suggested that, for regular structures, both elastic and plastic displacements be determined at an effective height of $0{\cdot}64H$. It is also suggested that the displaced plastic shape be considered to vary linearly from profile 1 to profile 2 as n increases from 4 to 20. The plastic displacement at $0{\cdot}64H$ can therefore be shown to be

for

$$n \le 4 \quad \Delta_{\mathrm{p}} = 0{\cdot}64\theta_{\mathrm{p}}H$$
$$n \ge 20 \quad \Delta_{\mathrm{p}} = 0{\cdot}44\theta_{\mathrm{p}}H$$
$$4 < n < 20 \quad \Delta_{\mathrm{p}} = [0{\cdot}64 - 0{\cdot}0125(n-4)]\theta_{\mathrm{p}}H \qquad (6.11)$$

For the column-sway mechanism (profile 3), h_{eff} should reflect the ductility level. Thus, approximately

$$h_{\mathrm{eff}} = [0{\cdot}64 - 0{\cdot}14(\mu_{\mathrm{s}} - 1)/\mu_{\mathrm{s}}]H \qquad (6.12)$$

The plastic displacement Δ_{p} is given, for a structure of n equal storey heights h_{s}, as

$$\Delta_{\mathrm{p}} = \theta_{\mathrm{p}}h_{\mathrm{s}}$$
$$\text{i.e.} \quad \Delta_{\mathrm{p}} = \theta_{\mathrm{p}}H/n \qquad (6.13)$$

Calculating the structural yield displacement Δ_{y} at the effective height h_{eff}, the ultimate displacement capacity is given by

$$\Delta_{\mathrm{u}} = \Delta_{\mathrm{y}} + \Delta_{\mathrm{p}}$$

and the displacement ductility by

$$\mu_{\mathrm{s}} = \Delta_{\mathrm{u}}/\Delta_{\mathrm{y}}$$

In calculating the yield displacement, it is essential that member stiffness should include the influence of cracking, and that foundation compliance effects should be considered. Ideally, member stiffness should be based on moment–curvature analyses, at conditions corresponding to first yield, taking into account the influence of axial load and longitudinal reinforcement ratios.

Failing this, simplifying recommendations, such as those given by Paulay and Priestley (1992) can be adopted with reasonable accuracy.

The equivalent viscous damping and effective stiffness can therefore be calculated as described with reference to Fig. 6.3, and the displacement-based assessment procedure carried out as described in section 6.3.2.

Any premature failure of members, for example due to inadequate shear strength, should be considered in the evaluation of the effective mechanism and in correct estimation of the available displacement.

Similarly to the case of the detection of a storey-sway mechanism, the problem of the deterministic evaluation of two different situations, characterized by completely different responses in terms of displacement capacity but separated by an infinitely small variation of some parameter, may arise. It will therefore be necessary to establish appropriate relative values of protection factors to assess particularly undesired events or when a potentially large scatter around the most probable value is anticipated.

Only after extensive application of the method, evaluation of the results of in-depth probabilistic studies, and systematic comparison with field or experimental or highly reliable numerical evidence, will the appropriate values of the protection factors be confidently evaluated.

6.5. Member strength and deformation capacity

In either a force-based or a displacement-based assessment process, the first and key step is the evaluation of the strength and deformation capacity of the critical sections and elements.

Actually, the approach outlined so far has assumed a knowledge of member strength and deformation so that storey shear force and drift capacity can be calculated. Some discussion of these points is warranted.

The basis of assessment should be to obtain a 'best estimate' of performance. Hence, it is inappropriate to use design values for material strength, which will generally be specified minima strength values, or at best, lower 5 percentile values.

On the contrary, it is not uncommon (e.g. EC8/1-4, 1994) to recommend reduced values for the concrete strength, to consider possible deterioration processes which may have taken place over time. This may result in assessing a different failure mode, with unpredictable consequences. As pointed out at the end of the previous section, it would be more sensible to apply correction factors to more brittle mechanisms, to ensure that the assessed displacement capacity will be available.

The material strengths should therefore be estimated at their most probable value, for example as follows.

(*a*) *Reinforcement*. If mill certificates are available, use the average for the appropriate bar size. Otherwise, adopt a value of $f_{ya} = 1 \cdot 1 f_y$ as the assessment yield strength, where f_y is the nominal yield strength.

(*b*) *Concrete*. There is likely to have been considerable strength increase with age since construction. Also, 28-day strength 'as built' is likely to have significantly exceeded the nominal value as a result of conservative mix design. Since shear strength will often be of critical importance in seismic assessment, it is important to have an estimate of concrete strength which is as realistic as possible. The difference between using actual (f'_{ca}) and specified (f'_c) concrete strengths will frequently change the predicted failure mode from ductile flexure to brittle shear. Recent tests on concrete in 30-year old bridges in California have consistently resulted in compression strengths of approximately twice nominal strength. Wherever possible, cores should be taken to assess typical strengths, or at least impact hammer measurements used and correlated with a few reference cores to determine f'_{ca}. Failing this, a strength of $f'_{ca} = 1{\cdot}5f'_c$ could be conservatively adopted, where f'_c is the nominal design value, unless visual inspection indicates poor quality or decayed concrete. In such cases, extensive testing based on a combination of ultrasonic and impact hammer tests should be considered mandatory, together with some testing of cores to validate the non-destructive test results.

6.5.1. Beam hinge flexural capacity

Beam flexural strength should be assessed using an ultimate compression strain of 0·005, since this represents a conservative estimate of the onset of concrete damage. Actually, a lower limit on recorded crushing strains of plastic hinges forming against supporting members (in this case, columns) for normal strength concrete ($f'_{ca} < 50$ MPa) can be estimated around 0·006 (Mander *et al.*, 1988). However, if a moment–curvature analysis incorporating strain-hardening of reinforcement is used to assess flexural strength, positive moment capacity predicted at an extreme fibre compression strain of 0·004 may correspond to excessive tensile strains, when, as is normally the case, the area of top (compression) reinforcement exceeds that of the bottom (tension) reinforcement. For such cases, flexural strength should be assessed when peak tensile strain in reinforcement is about $\epsilon_s = 0{\cdot}02$. The effective slab width contributing to beam negative moment capacity is primarily a function of ductility level and slab reinforcement details. As an approximation guide, the width may be taken as the lesser of one-quarter of the beam span, or one-half of the transverse beam span for a beam framing into an interior column, or 50% of this value for beams framing into an exterior column. These values may be conservatively low at high ductility levels. A more complete consideration of these effects is given by Paulay and Priestley (1992).

6.5.2. Plastic rotation capacity

The plastic rotation capacity of the beam plastic hinges defines the plastic storey drift θ_p in a beam-sway mechanism. This will depend primarily on the detailing of the transverse reinforcement in the potential plastic hinge regions at the beam ends. Priestley and Calvi (1991) proposed rather crude rules relating structural ductility capacity μ_s to level of detailing provided, with values of $\mu_s = 2$ corresponding to 'bad' detailing and $\mu_s = 6$ corresponding to detailing conforming to requirements of current codes, such as NZS3101 (1995). Interpolation was suggested for intermediate cases. In view of the variability of reinforcement detailing, and in relationships between storey drift and structural displacement ductility, this is felt to be too coarse. Although the precision with which plastic drift capacity of existing structures can be predicted is still not high, some more specific guidance can be given. In particular, the emphasis in the displacement-based assessment procedure must be on a specific quantification of plastic hinge rotation θ_p.

Figure 6.12 presents information relative to predicting θ_p for the beams of typical frames. Using techniques normally employed for assessment of bridge bents (Priestley *et al.*, 1996), the plastic rotation capacity can be expressed as

$$\theta_p = (\phi_u - \phi_y)L_p \tag{6.14}$$

where ϕ_u and ϕ_y are ultimate and yield curvatures, and L_p is the equivalent plastic hinge length, given by

$$L_p = 0{\cdot}08L + 0{\cdot}022f_{ya}d_{bl} \tag{6.15}$$

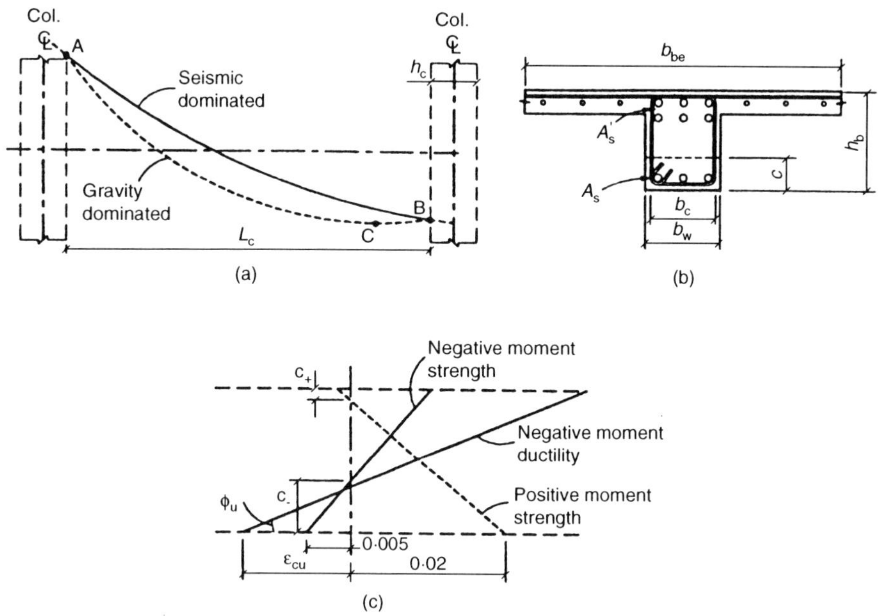

Fig. 6.12. Considerations for beam plastic hinges: (a) span elevation; (b) beam section; (c) strain profile

The first term in equation (6.15) represents the spread of plasticity due to member length, with L being the distance from the critical section to the point of contraflexure. The second term represents strain penetration into the supporting member (i.e. the column), with f_{ya} being the yield strength (MPa) of the beam longitudinal bars, of diameter d_{bl}. As shown in Fig. 6.12(a), the distance between the critical section and the point of contraflexure will depend on relative flexural strength of positive moment and negative moment plastic hinges, and the relative importance of seismic and gravity moments. However, it is suggested that for negative moment plastic hinges, which will generally form against the column face, (point A, Fig. 6.12(a)) a length $L = 0{\cdot}5L_c$, where L_c = beam clear span, be assumed. This is a reasonable reflection of the fact that (*a*) negative moment capacity will exceed positive moment capacity, and (*b*) high shear stress levels in the plastic hinge region will tend to extend the effective plastic hinge length due to tension shift effects.

The positive moment hinge could form at the column face, (point B, Fig. 6.12(a)) or within the span (point C), depending on the influence of gravity loads on the beam. However, the location, which is always hard to define due to uncertainty as to the magnitude of gravity loads, and the plastic rotation capacity of the positive moment hinge are of little interest in assessment because the plastic rotation will generally greatly exceed the rotational capacity of the negative-moment hinge. This is a consequence of top reinforcement area A_s' (including slab contribution) exceeding bottom reinforcement area A_s, and effective compression zone width b_{bc} for positive moments exceeding the web width b_w, appropriate for negative moments (see Fig. 6.12(b) and (c)). This results in a greatly reduced compression zone depth c_+ for positive moments compared to that for negative moments c_- as illustrated in Fig. 6.12(c). Since compatibility of the storey deformed shape requires that the plastic rotations of all plastic hinges along a beam are essentially equal at any given stage of response, and since plastic hinge lengths for positive moment can be expected to exceed those for negative moment, it follows that the critical condition, corresponding to attaining the ultimate compression strain ϵ_{cu} in a plastic hinge will always be in a negative-moment hinge. It can readily be shown that the theoretically feasible condition of attaining ultimate tensile strain in the positive-moment hinge is unrealistic at curvatures corresponding to the ultimate negative-moment curvature.

Figure 6.12(c) shows strain conditions to be used for estimating flexural strength of positive- and negative-moment hinges, and ultimate conditions for the negative-moment hinge. For 'unconfined' conditions, corresponding to

- only corner bars restrained against buckling by a bend of transverse reinforcement
- hoop stirrup ends not bent back into the core
- spacing of hoop or stirrup sets in the potential plastic hinge, such that

$$s \geq d/2$$

or

$$s \geq 16d_{\mathrm{bl}},$$

the ultimate concrete strain should be assumed to be 0·005, thus corresponding to conditions at determination of flexural strength.

For 'fully confined' conditions, corresponding to details satisfying current codes, i.e.

- all beam bars in the lower layer (i.e. if more than one) of bottom reinforcement restrained against buckling by transverse reinforcement of diameter greater than $d_{\mathrm{bl}}/4$
- all transverse reinforcement anchored by hooks bent back into the core by standard 135° hooks or equivalent anchorages
- spacing of hoop or stirrup sets not less than $s = d/4$ or $s = 6d_{\mathrm{bl}}$,

the ultimate concrete strain may be assumed to be

$$\epsilon_{\mathrm{cu}} = 0{\cdot}004 + \frac{1{\cdot}4\rho_{\mathrm{s}} f_{\mathrm{yh}} \epsilon_{\mathrm{su}}}{f'_{\mathrm{cc}}} \tag{6.16}$$

where the volumetric ratio of transverse reinforcement ρ_{s}, may be approximated as

$$\rho_{\mathrm{s}} = 1{\cdot}5A_{\mathrm{v}}/b_{\mathrm{c}}s \tag{6.17}$$

where A_{v} = total area of transverse reinforcement in a layer, at spacing s, and b_{c} = width of beam core, measured from centre to centre of the peripheral transverse reinforcement in the web. In equation (6.16) f_{yh} is the yield strength of the transverse reinforcement, ϵ_{su} is the strain in the transverse steel at maximum stress, and f'_{cc} is the compression strength of the confined concrete. For older designs, it is recommended that $\epsilon_{\mathrm{su}} = 0{\cdot}15$ and 0·10 for f_{y} = 250–400 MPa and 400–500 MPa transverse reinforcements, respectively. As already discussed, in lieu of a more accurate analysis $f'_{\mathrm{cc}} = 1{\cdot}5f'_{\mathrm{cu}}$ may be assumed.

For intermediate conditions between unconfined and fully confined, interpolation will be required.

An example of this approach is given in Fig. 6.13, where moment–curvature curves for positive and negative moment bending of a typical beam section are shown. A bay length of 6 m is assumed, which, with a column size of 450 mm square, gives an effective clear span of 5·55 m. Top steel area including the contribution of slab reinforcement over a 3000 mm effective width is more than double the bottom steel area. Despite this high steel ratio, the strengths of the section in positive and negative bending are not greatly different at high curvatures, due to cover spalling and a deep compression zone depth for negative moments, and strain hardening for

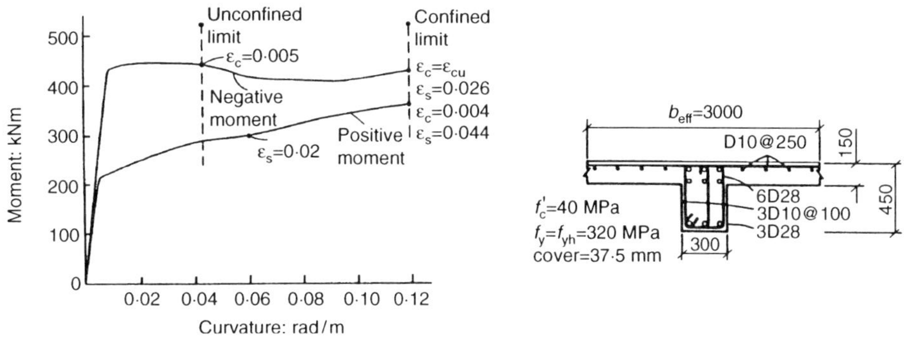

Fig. 6.13. Moment–curvature relationships for beam example

positive moments. At the limit curvature for unconfined negative bending ($\epsilon_c = 0{\cdot}005$) the positive moment hinge has a maximum extreme fibre strain of less than 0·0015, even assuming a reduced effective compression zone width of 1000 mm. If the longitudinal reinforcement is properly confined by sets of three Φ10 bars at 100 mm centres, the ultimate negative-moment curvature increases from 0·046 rad/m to 0·12 rad/m. At this curvature, spalling of cover concrete for the positive moment hinge is still not expected.

In fact, of course, the analysis for positive-moment bending is simplistic, since under cyclic loading, the bottom reinforcement will be unable to yield the top reinforcement in compression, and thus a steel couple will develop, with slightly reduced moment capacity. Nevertheless, the conclusion that positive-moment bending is not critical, remains.

For the example of Fig. 6.13, an effective plastic hinge length of $L_p = 0{\cdot}08 \times 2775 + 0{\cdot}022 \times 320 \times 28 = 419$ mm. is predicted from equation (6.15). With a yield curvature of $\phi_y = 0{\cdot}009$ rad/m (from moment–curvature analysis, or hand analyses), the plastic rotation capacity of the plastic hinge is found to be, for the unconfined case, $\theta_p = (0{\cdot}046 - 0{\cdot}009) \times 0{\cdot}419 = 0{\cdot}0155$ rad.

6.5.3. Column hinge flexural capacity

The procedure outlined above also applies, with minor changes, to hinges forming at column bases, or in column-sway inelastic mechanisms. However, the approximation for the volumetric ratio of transverse reinforcement in equation (6.17) should be replaced by a first principles approach. In fact, it will often be found that columns in older reinforced concrete frames have only nominal transverse reinforcement, and thus must be considered to be unconfined. Together with reduced plastic hinge length as a consequence of reduced member height compared with beam length, and reduced ultimate curvature as a consequence of axial compression, column plastic rotation capacity will generally be less than values estimated for beams, and values less than $\theta_p = 0{\cdot}01$ rad will be common.

Since axial load critically affects the ultimate curvature, it is essential that seismic axial forces be included when estimating column plastic rotation. The critical column will be the one with highest axial compression. Moment–curvature analyses will show that, while yield curvature is not greatly affected by axial load level, particularly when yield curvature is expressed in terms of equivalent elasto-plastic response, ultimate curvature, and hence plastic rotation capacity, is strongly dependent on axial load. This is illustrated in Fig. 6.14, where an unconfined end column of a frame, with nominal axial load of $P = 0{\cdot}2f_{ca}A_g$ is subjected to seismic axial force variations of $P_E = \pm 0{\cdot}2f_{ca}A_g$. The yield curvatures differ by less than 10% from the mean, while the ultimate curvatures at $P = 0$ and $P = 0{\cdot}4f_{ca}A_g$ are 263% and 61% of the value at $P = 0{\cdot}2f_{ca}A_g$, respectively.

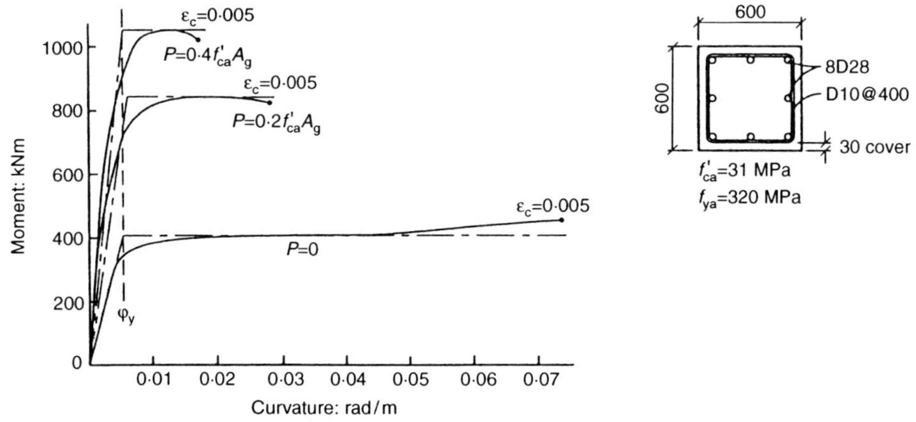

Fig. 6.14. Moment–curvature response of unconfined columns

6.5.4. *Column shear strength*

Existing concrete frames of the 1930–75 era will often be found to have only nominal transverse reinforcement, in the form of peripheral ties at spacing similar to the column dimensions. Shear strength assessment, using typical code equations for shear strength will frequently show these columns to be severely deficient, even when realistic, rather than nominal material strengths are used.

Code equations for shear strength are known to be excessively conservative, in many cases, and to show wide scatter when used to predict test results. It is therefore recommended that column shear strength be assessed using equations that are more realistic than those currently incorporated in codes, and which reflect the dependence of shear strength on flexural ductility. A recently developed model (Priestley *et al.*, 1994) which provides close agreement with experiments, expresses the nominal shear strength of columns as the sum of components due to concrete contribution (V_c), transverse reinforcement (V_s) and axial load V_p. Thus

$$V_n = V_c + V_s + V_p \tag{6.18}$$

where

$$V_c = 0{\cdot}8 A_{\text{gross}} k \sqrt{f'_{ca}} \tag{6.19}$$

$$V_s = \frac{A_v f_{yh}(D-c)}{s}\cot\theta \qquad \text{rectangular sections} \tag{6.20a}$$

$$V_s = \frac{\pi}{2}\frac{A_{sp} f_y (D-c)}{s}\cot\theta \qquad \text{circular sections} \tag{6.20b}$$

and

$$V_p = P\tan\alpha \tag{6.21}$$

Figure 6.15 describes the degradation of k in equation (6.19), with increasing curvature ductility $\mu_\phi = \phi/\phi_y$, and the meaning of tan α in equation (6.21). In equation (6.20), D is the section depth or diameter, c the neutral axis depth and θ is the angle of the critical inclined flexure shear cracking to the column axis, recommended to be taken as 30°. A major difference from other models is that the contribution of axial force to shear strength is represented by the horizontal component of the axial force strut formed between top and bottom of the column. A full description of the model for shear strength is available in Priestley *et al.* (1994) and in Benzoni *et al.* (1996).

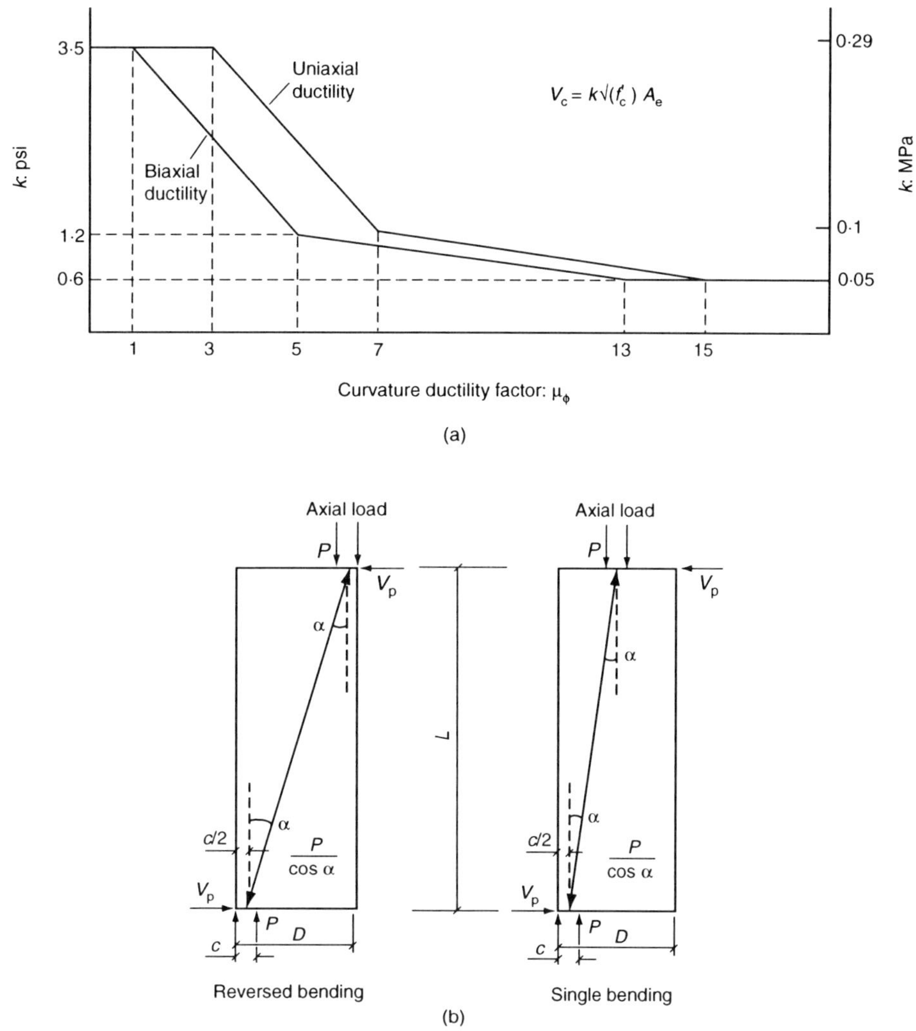

Fig. 6.15. Analysis for shear strength of columns: (a) degradation of concrete shear strength with ductility; (b) contribution of axial force to column shear strength

The shear strength given by equation (6.18) is therefore a function of the curvature ductility at the member critical section. Two limiting values, V_{ni} and V_{nd} correspond to initial strength ($\mu_\phi \leq 1$) and residual strength ($\mu_\phi \geq 15$), respectively. If the shear V_{if} corresponding to initial flexural strength exceeds V_{ni}, then brittle shear failure occurs, and $\mu_\phi \leq 1$. If $V_{if} < V_{nd}$, then ductile flexural response is assured, and the maximum section ductility is limited by the flexural confinement details. When $V_{nd} < V_{if} < V_{ni}$, then a shear failure is expected at some limited ductility. Interpolation using equation (6.18) and Fig. 6.15(a) enables the section curvature $\mu_{\phi S}$ to be calculated at which shear failure is expected.

Prediction of whether shear failure or flexural failure will occur, and the final section curvature ductility thus depends on relative magnitude of ideal flexural and shear strength, which will often be based on assumed material properties, in the absence of appropriate test data. Variations in material properties — particularly in the yield strength of the longitudinal reinforcement — can make significant differences to the outcome of this assessment. This is particularly the case when the shear demand at ideal flexural strength V_{if} is only slightly lower than the ductile shear strength V_{nd}. In this case, a small increase in longitudinal yield strength can convert a predicted ductile response into a shear failure at significantly reduced ductility. As a consequence, and also to allow for scatter in the ratio between predicted and actual shear strength, it is recommended that equation (6.18) be used in conjunction with a shear strength reduction factor of $\phi_S = 0{\cdot}75$, when comparing shear strength and shear at nominal flexural strength of columns.

6.5.5. *Beam shear strength*

The model in Fig. 6.15(a) for shear strength degradation in plastic hinge regions has been extensively investigated for column sections. Although it is known that beam shear strength in plastic hinges reduces with ductility, an equivalent relationship has not been established for beams. In fact, there seems to be surprisingly little relevant data on which to base such a model. It would seem that there should not be much conceptual difference between a beam and a column with zero axial load and that, hence, equations (6.19) and (6.20) should also apply directly to beams. However, columns generally have distributed longitudinal reinforcement, ensuring the existence of a flexural compression zone under cyclic inelastic response, even when $P = 0$. With a beam section, the tension reinforcement area A_S may be less than the compression reinforcement area A'_s, and under inelastic response no concrete compression will occur. Under these circumstances, $V_c = 0$ would seem appropriate at high ductilities. On the other hand, in building frames, the condition of $A'_s > A_s$ is not likely to be critical for shear, since this represents the case of positive moment, when seismic and gravity shear forces act in opposition.

Until further test data become available, it is recommended that the shear strength of a negative-moment plastic hinge be assessed using equation (6.18), but with $k = 0{\cdot}2$ for $\mu_\phi \leq 3$ and $k = 0{\cdot}05$ for $\mu_\phi \geq 7$, in equation (6.19).

6.5.6. *Beam–column joint shear*

Current seismic design philosophy requires that considerable amounts of transverse reinforcement be placed in beam–column joints of reinforced concrete frames to assist in joint shear transfer (Paulay and Priestley, 1992, NZS3101, 1994, EC8, 1994). However, the design methods, while being quite prescriptive, do not provide the necessary information to assess the strength and deformation capacity of sub-standard joints. Since older frame structures were almost always constructed without special joint reinforcement, this is a cause for concern.

Recently, a considerable amount of research has been carried out in order to better quantify the performance of poorly designed joints. The vast body of data assembled by Japanese researchers (Kurase, 1987, Hakuto *et al.*, 1995) has been particularly useful in this regard. It is clear from this and other research carried out in New Zealand and the USA, that there is a significant difference to be expected in the seismic performance of interior and exterior joints. Some observations from a review of this research, and tentative recommendations, are made below.

The distinction between ‘interior’ and ‘exterior’ joints, as discussed herein, is clarified by Fig. 6.16. The joint in Fig. 6.16(a), part of an exterior frame, is considered an interior joint for loading in the plane of the frame, but is an exterior joint in the orthogonal direction, suggested by the dashed line. ‘True’ interior joints, such as shown in Fig. 6.17(b), may be subjected to seismic response as interior joints in either, or both of the two orthogonal directions. The corner joint of Fig. 6.16(c) deserves special attention. There is reason to expect that corner joints might represent the critical conditions in building frames because of the biaxial input, typically difficult reinforcement detailing problems involved in anchoring two orthogonal sets of beam bars in the joint, and the influence of variable axial load. Despite this concern, there are almost no test data available for this type of joint.

6.5.7. Interior joints

Review of the test data for interior joints indicates the following trends.

6.5.7.1. With lightly reinforced beams, or with columns with high axial force levels, joint cracking may not develop. The critical parameter here is the principal tension stress in the joint, rather than the shear stress level. Based on gross joint dimensions, a critical tension stress of

$$p_t = 0{\cdot}29\sqrt{f'_{ca}}$$

seems appropriate, where

$$p_t = \frac{f_a}{2} - \sqrt{\left[\left(\frac{f_a}{2}\right)^2 + \nu_j^2\right]} \tag{6.22}$$

and $f_a = P/A_{col}$, $\nu_j = V_{jh}/A_{col}$ are the average axial stress and shear stress in the joint core.

6.5.7.2. Beam–column joints with high shear stress levels tend to fail in shear regardless of the amount of transverse reinforcement. This is recognized in the USA by a limit of $\nu < 1{\cdot}7\sqrt{(f_{ca})}$MPa on joint shear stress. However, the reason for failure is the principal compression stress, and it is thus more logical to limit this directly, rather than through the shear stress, which does not recognize the influence of axial compression. Note that tests on joints with high axial loads and apparently adequate transverse reinforcement (Pessiki *et al.*, 1990, Beckingsale, 1980) and with the comparatively low shear stress levels of $\nu_c \approx 1{\cdot}0\sqrt{f'_{ca}}$ failed in shear. In both cases, the principal compression stress was about $0{\cdot}5f'_{ca}$. Thus a tentative upper limit for shear stress would be related to the principal compression stress, by

$$p_c = \frac{f_a}{2} + \sqrt{\left[\left(\frac{f_a}{2}\right)^2 + \nu_j^2\right]} \leq 0{\cdot}5f'_c \tag{6.23}$$

Inverting equation (6.23) yields

$$\nu_j \leq p_c\sqrt{\left[1 - \left(\frac{f_a}{p_c}\right)\right]} \tag{6.24}$$

where, for one-way joints, $p_c = 0{\cdot}5f'_{ca}$, as above, and for two-way joints, $p_c \approx 0{\cdot}45f'_{ca}$, to allow for effects of biaxial joint shear.

6.5.7.3. For beam–column joints with principal tension stress $p_t > 0{\cdot}29\sqrt{(f'_{ca})}$ and principal compression stress $p_c < 0{\cdot}5f'_{ca}$ (or $0{\cdot}45f'_{ca}$ for biaxial bending), failure may be due to joint shear, bond slip of rebar through the joint or beam flexural ductility. In virtually all cases of interior test joints within this range, the flexural strength of the beams on both sides of the joint was developed before joint failure occurred, unless the columns were weaker than the beams.

For this category of interior joint, Japanese test results lead to the following conclusions.

- The role of transverse reinforcement seems different for cases where beam bar slip occurs, and where it is restrained. In the former case, hoop strains are largely independent of the amount of hoop reinforcement. In the latter case, hoop strains decrease as amount of hoop reinforcement increases.

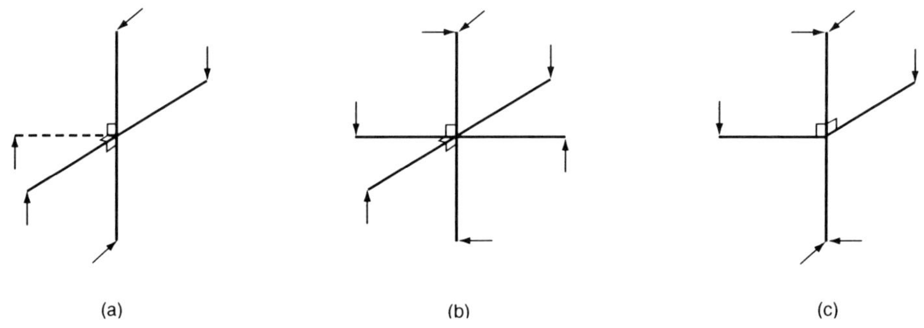

Fig. 6.16. Categories of beam-column joints: (a) one-way joint; (b) two-way joint; (c) corner joint

- When beam bar bond failure occurs, the hysteresis loops become very pinched. Strength degradation typically starts at a drift angle of about 2%.
- When beam bar bond slip through the joint is inhibited (by provision of a larger joint width–beam bar diameter ratio), more force is transmitted to the joint by bond. Failure initiates in the joint at drift angles of about 2%, if transverse reinforcement is insufficient to carry about 50% of the joint shear, and degrades more rapidly than when bond slip develops.

Definitive failure models are not yet available. Recently, however, Hakuto, Park and Tanaka (1995) have suggested the principal tension model of Fig. 6.17(a). It is seen that the lower limit of this is similar to the principal tension stress suggested in relation to equation (6.22). It seems that the upper limit, being based on tension rather than compression stress, may result in anomalies when either high or low axial compressions are present. An alternative, but similar, formulation based on principal compression stress and the observations listed above is suggested in Fig. 6.17(b). In this, the principal compression stress ratio p_c/f'_{ca} is related to plastic drift, rather than joint displacement ductility. Line 1 indicates response of a joint with low principal compression stress, for which joint failure is not predicted. Line 2 represents a case with higher principal compression stress. For this case, strength begins to decline once the failure surface has been reached.

A third, and simpler formulation is suggested in Fig. 6.17(c), where, provided that $p_c < 0{\cdot}5f'_{ca}$ (or $0{\cdot}45f'_{ca}$ for two-way joints), the joint shear strength ratio V_j/V_{jf} is assumed to start degrading at 1% drift, regardless of the actual shear stress or principal stress level. In this formulation, V_{jf} is the joint shear corresponding to beam flexural strength. This model has the merit of capturing the essentials of the failure mechanism noted by others for under-reinforced joints, but it should be stressed that neither the model of Fig. 6.17(c) nor that of Fig. 6.17(b) have been adequately tested to date. It is felt, however, that the model of Fig. 6.17(c) is likely to prove of adequate conservatism for assessment.

6.5.8. Exterior and corner joints

As with interior joints, a principal tension stress of about $0{\cdot}29\sqrt{(f'_{ca})}$MPa appears to be a lower limit for joint cracking. When beam reinforcement is anchored by bending away from the joint (see Fig. 6.18(a)), diagonal struts in the joint cannot be stabilized, and joint failure occurs at an early stage. The situation when beam reinforcement is bent down into the joint is illustrated in Figs 6.18(b) and 6.18(c). Joint cracking will first develop under positive beam moments, since axial force on the column is reduced for this direction of response. In a multi-storey building, the axial force variations in exterior, and in particular, in corner columns can be very high,

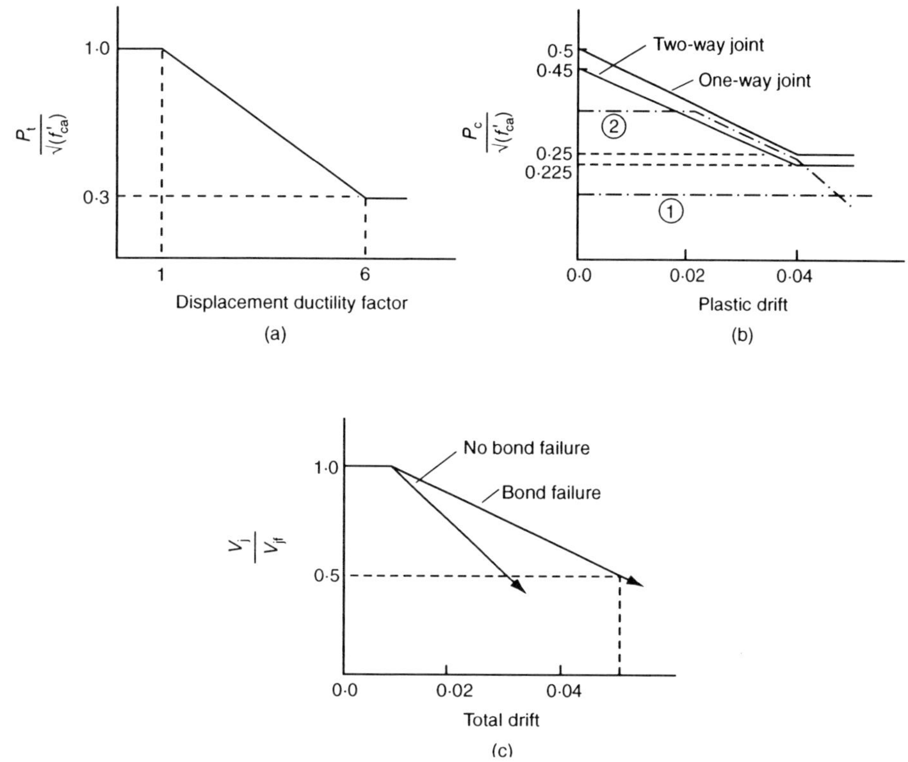

Fig. 6.17. Possible failure models for interior beam–column joints: (a) principal tension model of Hakuto, Park and Tanaka; (b) suggested principal compression model; (c) possible shear strength model

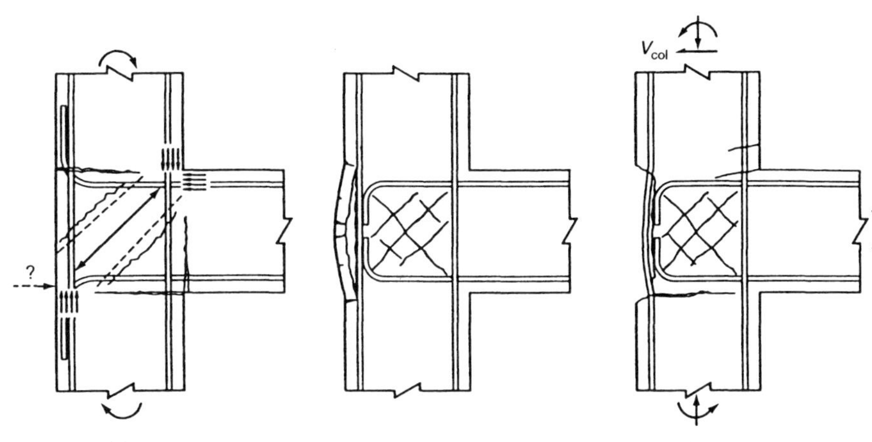

Fig. 6.18. Breakdown of unreinforced exterior joints: (a) beam bars bent away from joint; (b) beam bars bent in — cover cracking at back of joint; (c) loss of joint integrity

and as a consequence, cracking under negative moment will be delayed, and may not occur at all.

When cracking occurs, the joint tends to dilate horizontally. This places the cover concrete at the back of the joint in curvature and vertical cracking occurs on the weak plane at the line of column reinforcement, particularly if beam reinforcement hooks lie in the same plane. This is illustrated in Fig. 6.18(b). The cover concrete is likely to spall off under the increased compression load corresponding to beam negative moment. In unreinforced joints this severely degrades the anchorage of the beam bar hooks, which is needed to equilibrate the diagonal strut in the joint. The combined action of resistance to this diagonal strut and the pulling force from the beam reinforcement tension force tends to open the hooks, as shown in Fig. 6.18(c), further degrading joint performance. Joint degradation is then comparatively rapid.

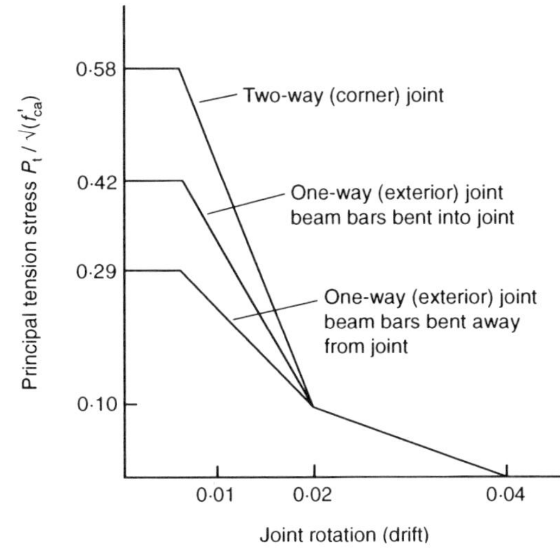

Fig. 6.19. Suggested strength degradation model for exterior and corner joints

Comparatively small amounts of transverse joint reinforcement greatly improve the behaviour. Joint dilation is reduced, as is the tendency for cover spalling at the back of the joint. This still occurs, but at a later stage of response. Straightening of the beam bar hooks is restrained, maintaining integrity of the diagonal compression strut.

It is evident that failure of exterior joints is primarily related to principal tension stress. The following tentative recommendations are based on tests of unreinforced exterior and corner joints.

(*a*) For beam bars bent away from the joint, joint failure can be considered to initiate at a principal tension stress of $0{\cdot}29\sqrt{(f'_{ca})}$ MPa.

(*b*) For beam bars bent down across the back of the joint, higher principal tension stresses are possible. The test data support principal tension stresses of $0{\cdot}42\sqrt{(f'_{ca})}$ MPa and $0{\cdot}58\sqrt{(f'_{ca})}$ MPa for exterior joints, and corner joints under biaxial response, respectively. Note that under diagonal response of corner joints the joint shear force is formed from vectorial addition of the orthogonal shears.

(*c*) Joint degradation after formation of cracking is governed by gradual reduction of the effective joint principal tension stress, in accordance with the relationship suggested in Fig. 6.19.

The information provided above enables an estimate to be made of the plastic storey drift that could occur when a joint-failure mechanism develops. Note that the degradation will often be found to be quite gradual in terms of storey shear strength reduction. However, if joint failure can occur at two adjacent levels of a building, a soft-storey sway mechanism can develop, and the structural ductility capacity μ_s will be found to be low. Also, energy dissipation with a joint failure mechanism is less than for a beam- or column-sway mechanism. It is recommended that a flat 10% effective viscous damping be used in the displacement-based assessment procedure outlined in section 6.3.2.

A strength reduction factor of $\phi_s = 0{\cdot}75$, as for shear, should be used in conjunction with calculated joint shear strength, to account for potential overstrength of beam plastic hinges, and variability of joint shear strength.

6.6. Structural wall buildings

Many of the principles outlined above are directly applicable to buildings where the principal form of lateral resistance is structural walls. Traditionally, buildings with structural walls in both orthogonal directions have performed well in earthquakes (Wyllie *et al.*, 1986) even when the

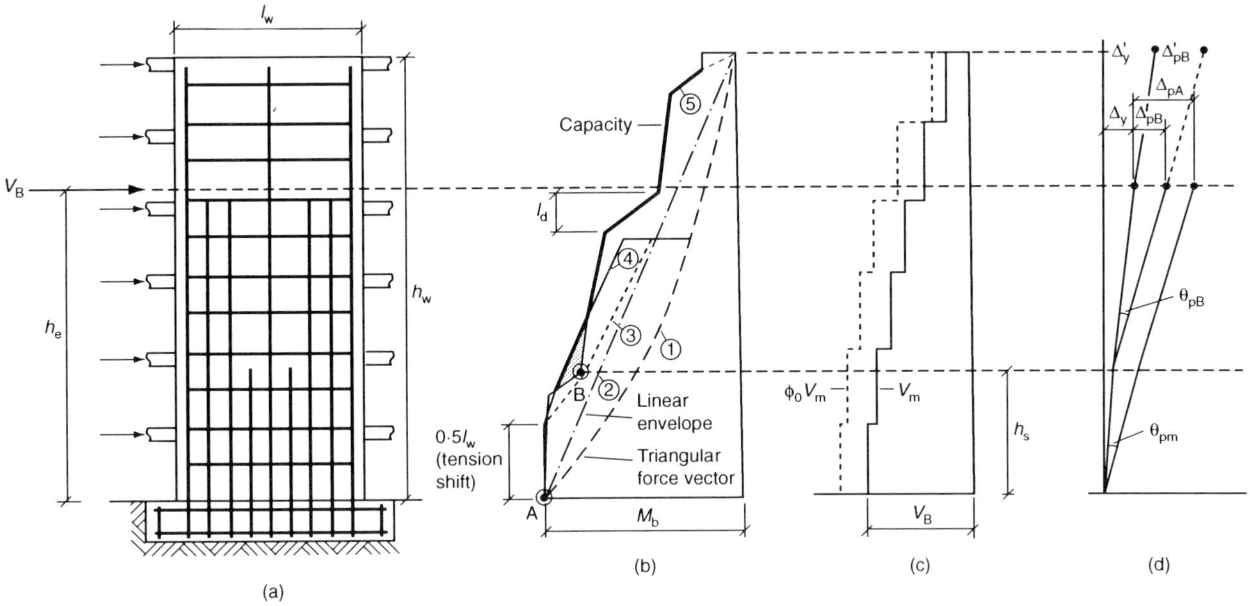

Fig. 6.20. Assessment of cantilever wall: (a) wall configuration; (b) moments at base capacity; (c) shears at base moment strength; (d) displacement profiles

walls have not been designed to capacity principles. However, when walls are provided in only one direction, and the frames in the orthogonal direction develop column-sway mechanisms, the excessive lateral drift in the walls can cause instability and premature collapse, as occurred in many buildings in the 1995 Kobe earthquake (Priestley *et al.*, 1995).

Some of the aspects of assessment of structural walls are illustrated in Fig. 6.20, which shows a simple cantilever wall, linked to other structural elements by flexible floor slabs. Flexural reinforcement (Fig. 6.20(a)) reduces with height, reflecting the anticipated decreased flexural demand. Fig. 6.20(b) compares various moment profiles up the wall height with the computed moment capacity. Profile 1 assumes a base moment equal to calculated flexural strength, and an inverted triangle distribution of lateral force. The effects of higher modes would be included by using profile 2 for design, where moments decrease linearly with height to zero (Paulay and Priestley, 1992, NZS3101, 1995). To account for effects of diagonal cracking, a tension shift is normally applied to the moment diagram, displacing profile 2 vertically by an amount equal to the wall length l_w.

For assessment, this approach, which is desirably conservative for design, can be somewhat modified. First, the tension shift can be reduced to $0{\cdot}5l_w$, as shown in Fig. 6.20(b), where profiles 3 and 4 show tension shift applied to profiles 1 and 2 respectively. Secondly, if the wall nominal shear stress $V/A_w < 0{\cdot}2\sqrt{(f'_{ca})}$ MPa, diagonal cracking is unlikely to develop and tension shift is inappropriate. This critical shear stress level is achieved slightly above wall mid-height in the example given in Fig. 6.20. Above this level, profiles 3 and 4 revert to profiles 1 and 2.

The calculated moment capacity is represented by profile 5, which includes effects of gradual moment increase over the development length l_d, adjacent to rebar termination. A cracking moment capacity M_{cr} exceeds ultimate strength at the top of the wall.

Comparison of profiles 4 and 5 indicate that plastic hinging could develop at the wall base (point A), or just below the second floor, at point B. However, it is noted that the capacity at B exceeds profile 3, and thus hinging is not predicted under the normal triangular force vector, but only when higher modes are considered. It is unlikely that significant inelastic displacement demand will occur in this condition. Consequently, for assessment, it is decided to base inelastic response on a base hinge. If the capacity at B (or elsewhere) was less than that of profile 3, a different conclusion would be reached.

Assessment of shear strength is included in Fig. 6.20(c). Here, V_m is the shear force distribution corresponding to the triangular force vector, and base moment capacity M_b. The base shear force is

$$V_B = M_b/h_e \tag{6.25}$$

In capacity design for new buildings, the design shear force distribution would be taken as

$$V = \omega\phi_0 V_m \tag{6.26}$$

where ϕ_0 represents potential flexural overstrength due to high material strengths, and strain hardening, and ω represents dynamic shear amplification due to higher mode effects (Paulay and Priestley, 1992, NZS3101, 1995). It will be noted that assuming maximum values for ω and ϕ_0 simultaneously is conservative since maximum curvature ductility (and hence maximum ϕ_0) cannot occur simultaneously with maximum higher mode response. Also, dynamic analyses tend to indicate that the duration of dynamically amplified shears is typically extremely short. It therefore seems unlikely that sufficient energy could be fed into the wall to sustain the displacements necessary for a full shear failure. Consequently, it is suggested that the shear strength of the wall be checked only against $V = \phi_0 V_m$. To simplify the assessment, it is suggested that this be effected by use of an artificially low shear strength reduction factor $\phi_s = 0{\cdot}75$, as was suggested for frame members with shear demand corresponding to nominal flexural strength.

Wall shear strength may be assessed using equations (6.18) to (6.21), since the difference between walls and columns is primarily one of semantics. Agreement of predicted and measured shear strength, using these equations appears reasonable, although exhaustive testing has not yet been carried out.

Calculation of plastic displacement capacity also follows the methods suggested for frame members. For a hinge forming at the wall base, the plastic hinge length is given by equation (6.15), substituting $L = h_e$, and the plastic rotation by equation (6.14). An ultimate compression strain for unconfined concrete may again be conservatively taken as $\epsilon_c = 0{\cdot}005$, and the effect of confinement based on the volumetric ratio of confinement at the wall ends, using equation (6.16). In the event that a wall hinge is predicted at point B, Fig. 6.20(b), the plastic displacement at the centre of seismic force will be

$$\Delta_{pB} = \theta_{pB}(h_e - h_B) \tag{6.27a}$$

instead of

$$\Delta_{pA} = \theta_{pA} h_e \tag{6.27b}$$

However, if plastic hinging at B is a consequence of higher mode effects, as previously discussed, based on a liner moment–demand envelope, it is probably unrealistically conservative to carry out the displacement calculation at the effective height h_e, which is based on the assumption of a triangular force vector. Perhaps calculating Δ'_y and Δ'_{pB} at the wall top would be more appropriate.

The above discussion is simplistic, but illustrates most of the aspects that need to be considered in a wall assessment. For multiple cantilever walls, the force–deflection response of all walls can be added to provide the total response from which the structural effective stiffness and damping, in accordance with Fig. 6.4, are estimated. The assessment of coupled walls requires procedures which are intermediate between those for walls and frames, although no conceptual difficulties arise. In this context, it should be mentioned that the coupling effects of reinforced concrete slabs can be considerable in increasing elastic stiffness and lateral strength of structural wall buildings.

Finally, the capacity of the floor diaphragms, and their connections to the structural walls must be checked to ensure that they can distribute the inertia forces adequately in accordance with assumptions made in the assessment.

6.7. Conclusions

The authors of this Design Guide are well aware that the flavour of this chapter is quite different from that of the rest of the book. This is essentially due to the different maturity of the material presented in this chapter, which is based largely on the research work of members of the task group rather than on well-established state-of-the-art knowledge. Nevertheless, it was felt appropriate to include some discussion on the applicability of system-based, rather than member-based, approaches to the assessment of existing reinforced concrete structures.

The advantages of such an approach are actually more evident and decisive in the case of existing structures. In case of new design, the structure can be forced to respond with a particular mechanism, or, at least, undesired response modes can be excluded, applying appropriate capacity design principles, as discussed in the rest of this book. The system response mode can therefore be assumed as a design datum, and it may be possible to obtain similar results from simpler verification of single, critical, structural members. When dealing with an existing structure, the global response of the system cannot be determined without a careful consideration of the relative strength of sections, joints, members and storeys, it being impossible to assume at the beginning of the process.

It is interesting to note, even if this topic is outside the scope of this Design Guide, that re-design for retrofitting may be based on selective strengthening (or, paradoxically, weakening) of single element failure modes, with the objective of changing the post-elastic mechanism.

It has been discussed that displacement-based approaches seem to be better suited to a system-based assessment procedure than traditional force-based approaches. Even if more experimental and numerical research (i.e. extended comparisons between results obtained with simplified approaches and reliable non-linear dynamic models) is needed before drawing final conclusions and implementing displacement-based assessment procedures into common practice, it is felt that the superiority of displacement-based approaches for system assessment is somehow related to the intimate nature of strength, naturally oriented to local events and single critical elements, and of displacement, more immediately connected with storey drift and global structural response.

The quality of the results of an assessment process will depend largely on the reliability of the procedures used for the assessment of member strength. The discussion of these procedures comes last in this book, but first in a real application, as clearly pointed out in Fig. 6.7, where 'member strength' occupies the first box. In this respect, two different aspects of the procedure are discussed in this chapter. The first has a methodological relevance, and tries to clarify the logic of force- and displacement-based assessment procedures. The second one has a more practical relevance, presenting some up-to-date approaches to evaluating strength and deformation capacity of critical sections and elements. The importance of this last part is essentially due to the necessity of evaluating the response of underdesigned elements, for most of which insufficient research results are available.

The individuation of the correct post-elastic mechanism depends on the appropriate evaluation of the relative strength of different elements. Only on this basis is it reasonable to define a single-degree-of-freedom equivalent model. Still, the simulation of the real response using a single mode deformation pattern may, in some cases, penalize the quality of the predicted results, and the possibility of using several displacement shapes should be explored. In this respect, however, the multiple mode representation based on several elastic modes, commonly used in a force-based approach, is not necessarily superior.

References

Chapter 2

ACI (1992). *Building code requirements for reinforced concrete.* ACI 318–89, American Concrete Institute.

Architectural Institute of Japan (1990). *Design guidelines for earthquake resistant reinforced concrete buildings based on ultimate strength concept.* November 1990, 337 pp.

ATC–19 (1995). *Structural response modification factors.* Applied Technology Council.

CEN (1994). *European Prestandard Eurocode 8: design provisions for earthquake resistance of structures.* ENV 1998, 1–1 to 3, Brussels.

FEMA (1995). *NEHRP Recommended provisions for seismic regulations for new buildings.* (1994 Edition), FEMA 222A, May, *Part 1 — Provisions* (290 pp), *Part 2 — Commentary* (335 pp).

NIST (1992). *Assessment of the seismic provisions of model building codes.* National Institute of Standards and Technology GCR 91598 (114 pp).

NZ S 3101: 1995 *Concrete structures standard.* Standards New Zealand, Wellington, 1995.

NZ S 4203: 1992 *Code of practice for general structural design and design loadings for buildings.* Standards New Zealand, Wellington, 1992.

UBC (1994). *Uniform building code.* Vol. 2, Structural design provisions, International Conference of Building Officials (ICBO).

Chapter 3

Atalik T.S. and Utku S. (1976). Stochastic linearization of multi-degree-of-freedom non-linear systems. *Earth. Eng. Struct. Dyn.*, **4**, 411–20.

Baber T.T. (1984). Nonzero mean random vibration of hysteretic systems. *Am. Soc. Civ. Engrs, J. Engng Mech.*, **110**, 7, 1036–49.

Baber T.T. (1986). Nonzero mean random vibration of hysteretic frames. *Computers & Structures*, **23**, 2, 265–77.

Baber T.T. and Wen Y-K. (1981). Random vibration of hysteretic degrading systems. *Am. Soc. Civ. Engrs, J. Engng Mech. Div.*, **107**, EM6, 1069–87.

Colangelo F. (1994). LESS. A finite elements program for stochastic analysis of non-linear framed structures. DISAT Dept. Report 6, University of L'Aquila, Italy.

Colangelo F., Giannini R. and Pinto P.E. (1994). Stochastic linearization for the assessment of behavior factors and capacity design provisions in Eurocode 8. *Proc. 10th European Conf. on Earthq. Engng*, Wien, 947–52.

Colangelo F., Giannini R. and Pinto P.E. (1995). Calibration of capacity design coefficient for RC frames by means of stochastic linearization. *Proc. VII Nat. Conf. on Seism. Engng*, Siena, 973–82 (in Italian).

Colangelo F., Giannini R. and Pinto P.E. (1996). Seismic reliability analysis of RC structures with stochastic properties. *Struct. Safety*, **18**, 2/3, 151–68.

Eurocode 8 (1994). *Design provisions for earthquake resistance of structures.* ENV 1998-1-1/-3, CEN, Brussels.

Park Y.J., Wen Y-K. and Ang A. (1986). Two-dimensional random vibration of hysteretic structures. *Earth Engng Struct. Dyn.*, **14**, 543–57.

Pinto P.E., Colangelo F. and Giannini R. (1995). Stochastic linearization technique for the calibration of capacity design factors of RC frames. *Festschrift Prof. Bachmann — IBK Zurich* 96–102.

Sues R.H. and Wen Y-K. (1985). Stochastic evaluation of seismic structural performance. *Am. Soc. Civ. Engrs, J. Struct. Engng*, **111**, 6, 1204–18.

Wen Y-K. (1976). Method for random vibration of hysteretic systems. *Am. Soc. Civ. Engrs, Engng Mech. Div.*, **102**, EM2, 249–63.

Chapter 4

Ang A.H-S. (1989). Probabilistic seismic safety and damage assessments of structures (State-of-the-art report). *Proc. 9th World Conf. on Earthq. Engng*, Tokyo–Kyoto, Japan, Aug. 1988, **VIII**, 717–28.

Banon H. and Veneziano D. (1982). Seismic safety of reinforced concrete members and structures. *Earthq. Engng and Struct. Dynamics*, **10**, 10, 179–93.

Banon H., Biggs J.M. and Irvine H.M. (1981). Seismic damage in reinforced concrete frames, *Am. Soc. Civ. Engrs, J. Struct. Div.*, **107**, No. ST9, 1713–29.

Bertero V.V. and Bresler B. (1977). Design and engineering decision: failure criteria (limit states). *Developing methodologies for evaluating the earthquake safety of existing buildings*. Earthquake Engineering Research Centre, University of California, Berkeley, CA, Rep. EERC 77-06, Feb.

Blume, J.A., Newmark N.M. and Corning L.H. (1961). Design of multistorey reinforced concrete buildings for earthquake motions, PCA, Skokie, Illinois.

Bracci J.M. *et al.* (1989). Deterministic model for seismic damage evaluation of reinforced concrete structures, State University of New York at Buffalo, Techn. Rep. NCEER, 89-0033.

CEB TG III/6 (1994). Behaviour and analysis of reinforced concrete structures under alternate actions inducing inelastic response — Vol. 2: Frame members. CEB Bull. d'Information 220, Lausanne, Section 4.5.

Chai Y.H., Priestley M.J.N. and Seible F. (1994). Analytical model for steel-jacketed RC circular bridge columns, *Am. Soc. Civ. Engrs, J. Struct. Engng*, **120**, 8, 2358–76.

Chopra A.K. and Kan C. (1973). Effects of stiffness degradation on ductility requirements for multistorey buildings. *Earthq. Engng and Struct. Dynamics*, **2**, 1, 35–45.

Chung Y.S., Shinozuka M. and Meyer C. (1988). *Automated seismic design of reinforced concrete buildings*. State University of New York at Buffalo, Tech. Rep. NCEER 88-0024.

Chung Y.S., Meyer C. and Shinozuka M. (1987). *Seismic damage assessment of reinforced concrete members*. State University of New York at Buffalo, Tech. Rep. NCEER 87-0022.

Chung Y.S., Meyer C. and Shinozuka M. (1989). A new damage model for RC structures. *Proc. 9th World Conf. on Earthq. Engng*, Tokyo–Kyoto, Japan, Aug. 1988, **VII**, 205–10.

Chung Y.S., Meyer C. and Shinozuka M. (1989). Modeling of concrete damage, *ACI Struct. J.*, **86**, 3, May–June, 259–71.

Ciampoli M., Giannini R., Nuti C. and Pinto P.E. (1989). Seismic reliability of non-linear structures with stochastic parameters by directional simulation, *Proc. 5th Int. Conf. on Structural Safety and Reliability (-ICOSSAR 96)*, San Francisco, California, **II**, 1121–28.

Clough R.W., Benuska K.L. and Wilson E.L. (1965). Inelastic earthquake response of tall buildings, *Proc. 3rd World Conf. on Earthq. Engng*, Auckland and Wellington, New Zealand, **II**, 68–84.

DiPasquale E. and Cakmak A.S. (1987). *Detection and assessment of seismic structural damage*. State University of New York at Buffalo, Tech. Rep. NCEER 87-0015.

DiPasquale E. and Cakmak A.S. (1990). Seismic damage assessment using linear models. *Soil Dynamics and Earthq. Engng*, **9**, 4, 194–15.

DiPasquale E., Ju J-W., Askar A. and Cakmak A.S. (1990). Relation between global damage indices and local stiffness degradation. *Am. Soc. Civ. Engrs, J. Struct. Engng*, **116**, 5, 1440–56.

Dolce M., Kappos A., Zuccaro G. and Coburn A.W. (1995). Report of the EAEE WG3 — Vulnerability and risk analysis, *10th Europ. Conf. on Earthq. Engng*, Vienna, Austria, Aug.–Sept. 1994, **4**, 3049–77.

Fardis M.N. (1995). Damage measures and failure criteria for reinforced concrete members, *Proc. 10th Europ. Conf. on Earthq. Engng*, Vienna, Austria. Balkema, Rotterdam, **2**, 1377–82.

Garstka B., Krätzig W.B. and Stangenberg F. (1993). Damage assessment in cyclically loaded reinforced concrete members. *Structural dynamics — Eurodyn '93*. Balkema, Rotterdam, **1**, 121–28.

Giberson M.F. (1967). The response of nonlinear multi-storey structures subjected to earthquake excitation. California Institute of Technology, Pasadena, CA, PhD thesis.

Gunturi S.K.V. and Shah H.C. (1992). Building specific damage estimation. *Proc. 10th World Conf. on Earthq. Engng*, Madrid, Spain, **10**, 6001–06.

Hwang T.H. (1982). *Effects of variation in load history on cyclic response of concrete flexural members*. Dept. of Civil Engineering, University of Illinois, Urbana, Ill, PhD thesis.

Hwang H.T. and Scribner C.F. (1984). RC member cyclic response during various loadings. *Am. Soc. Civ. Engrs, J. Struct. Engng*, **110**, 3, 477–89.

Jeong G.D. and Iwan W.D. (1988). The effect of earthquake duration on the damage of structures. *Earthq. Engng and Struct. Dynamics*, **16**, 1201–11.

Kappos A.J. (1990). Sensitivity of calculated inelastic seismic response to input motion characteristics. *Proc. 4th US Nat. Conf. on Earthq. Engng*, Palm Springs, California, May 1990, EERI, Vol. 2, 25–34.

Kappos A.J. (1991). Analytical prediction of the collapse earthquake for RC buildings: suggested methodology. *Earthq. Engng and Struct. Dynamics*, **20**, 2, 167–76.

Kappos A.J. and Antoniades K. (1995). Seismic performance assessment of RC buildings designed to the 1994 Eurocode 8. *Proc. 5th SECED Conf. on European Seismic Design Practice*, Chester, UK, Oct. 1995, Balkema, Rotterdam, 349–57.

Kappos A.J. and Xenos A. (1996). A reassessment of ductility and energy-based seismic damage indices for reinforced concrete structures, *Proc. Eurodyn '96 (3rd Europ. Conf. on Struct. Dynamics)*, Florence, Italy, June 1996, **2**, 965–70.

Kappos A.J., Stylianidis K.C. and Michailidis C.N. (1996). A methodology for developing loss scenarios, with an application to the city of Thessaloniki, *Proc. (in CD ROM) 11th World Conf. on Earthq. Engng*, Acapulco, Mexico, June 1996, Paper No. 2057.

Kappos A.J., Stylianidis K.C. and Penelis G.G. (1991). Analytical prediction of the response of structures to future earthquakes, *European Earthq. Engng*, **5**, 1, 10–21.

Lybas J. and Sozen M.A. (1977). *Effect of beam strength and stiffness on dynamic behavior of reinforced concrete coupled walls*. University of Illinois, Urbana, IL, Civil Engineering Studies, Struct. Res. Series 444.

Mark K.M.S. (1976). Nonlinear dynamic response of reinforced concrete frames. Dept. Civil Engineering, Massachusetts Institute of Technology, Cambridge, MA, Res. Rep. R76-38.

Meyer I.F. Krätzig W.B., Stangenberg F. and Meskouris K. (1988). Damage prediction in reinforced concrete frames under seismic actions. *Europ. Earthq. Engng*, **3**, 1, 9–15.

Mizuhata K. and Maeda Y. (1989). Study on evaluation of cumulative damage earthquake response of structures based on computer-actuator on-line test. *Proc. 9th World Conf. on Earthq. Engng*, Tokyo–Kyoto, Japan, Aug. 1988, **IV**, 83–88.

Monti G. and Nuti C. (1992). Nonlinear cyclic behavior of reinforcing bars including buckling. *Am. Soc. Civ. Engrs, J. Struct. Engng*, **118**, 12, 3268–84.

Mork K.J. (1992). Stochastic analysis of reinforced concrete frames under seismic excitation. *Soil Dynamics and Earthq. Engng*, **11**, 3, 145–61.

Nielsen S.R.K., Köylüoglou H.U. and Cakmak A.S. (1992). One and two-dimensional maximum softening damage indicators for reinforced concrete structures under seismic excitation. *Soil Dynamics and Earthq. Engng*, **11**, 435–43.

Ogawa J. and Shiga T. (1989). Earthquake damage index for RC columns. *Proc. 9th World Conf. on Earthq. Engng*, Tokyo–Kyoto, Japan, Aug. 1988, **IV**, 425–50.

Oliveira C.S. (1977). Seismic risk analysis for a metropolitan area, *Proc. 6th World Conf. on Earthq. Engng*, Delhi, India, **I**, 686–94.

Papia M. and Russo G. (1989). Compressive concrete strain at buckling of longitudinal reinforcement. *Am. Soc. Civ. Engrs, J. Struct. Engng*, **115**, 2, 832–97.

Park Y-J., Ang A.H-S. and Wen Y.K. (1987). Damage-limiting aseismic design of buildings. *Earthquake Spectra*, **3**, 1, 1–25.

Park Y-J. and Ang A.H-S. (1985). Mechanistic seismic damage model for reinforced concrete. *Am. Soc. Civ. Engrs, J. Struct. Engng*, **111**, 4, 722–39.

Park Y-J., Reinhorn A.M. and Kunnath S.K. (1989). Seismic damage analysis of RC buildings. *Proc. 9th World Conf. on Earthq. Engng*, Tokyo–Kyoto, Japan, Aug. 1988, **VII**, 211–16.

Park Y-J., Reinhorn A.M. and Kunnath S.K. (1987). *IDARC: inelastic damage analysis of reinforced concrete frame–shear wall structures*. State University of New York at Buffalo, Tech. Rep. NCEER 87-008.

Paulay T. and Priestley M.J.N. (1992). *Seismic design of reinforced concrete and masonry buildings*. J. Wiley & Sons, New York.

Penelis G.G. and A.J. Kappos (1996). *Earthquake-resistant concrete structures*. E & FN Spon (Chapman & Hall), London.

Powell G.H. and Allahabadi R. (1988). Seismic damage prediction by deterministic methods: concepts and procedures. *Earthq. Engng and Struct. Dynamics*. **16**, 719–34.

Priestley M.J.N., Verma R. and Xiao Y. (1994). Seismic shear strength of reinforced concrete columns. *Am. Soc. Civ. Engrs, J. Struct. Engng*, **120**, 8, 2310–29.

Priestley M.J.N. (1993). *Seismic damage measures and member models*. Working document CEB TG III/2, May, 10 pp.

Reinhorn A.M. *et al.* (1988). Damage assessment of reinforced concrete structures in eastern United States. State University of New York at Buffalo, Tech. Rep. NCEER 88-0016.

Roufaiel M.S.L. and Meyer C. (1987). Analytical modeling of hysteretic behaviour of RC frames. *Am. Soc. Civ. Engrs, J. Struct. Engng*, **113**, 3, 429–44.

Saiidi M. and Sozen M.A. (1979). *Simple and complex models for nonlinear seismic response of reinforced concrete structures*. Civil Engineers Studies, University of Illinois, Urbana, **IL**, Struct. Res. Series 465.

Seidel M.J., Reinhorn A.M. and Park Y.J. (1989). Seismic damageability assessment of RC buildings in Eastern US. *Am. Soc. Civ. Engrs, J. Struct. Engng*, **115**, 9, 2184–2203.

Soleimani D., Popov E.P. and Bertero V.V. (1979). Nonlinear beam model for RC frame analysis. *Proc. 7th Conf. on Electronic Computation, Am. Soc. Civ. Engrs, Struct. Div.*, St Louis, Missouri, 483–509.

Sozen M.A. (1981). Review of earthquake response of RC buildings with a view to drift control. *State-of-the-art in earthquake engineering*. Kelaynak Press, Ankara, Turkey.

Stevens N.J., Uzumeri S.M. and Collins M.P. (1991). Reinforced concrete subjected to reversed cyclic shear — experiments and constitutive model. *ACI Struct. J.*, **84**, 2, 135–46.

Stone W.C. and Taylor A.W. (1994). ISDP: Integrated approach to seismic design of reinforced concrete structures. *Am. Soc. Civ. Engrs, J. Struct. Engng*, **120**, 12, 3548–66.

Takahashi J., Shibata A. and Shiga T. (1989). Crack indices of reinforced concrete shear walls for seismic damage evaluation. *Proc. 9th World Conf. on Earthq. Engng*, Tokyo–Kyoto, Japan, Aug. 1988, **IV**, 547–52.

Takizawa H. (1976). Notes on some basic problems in inelastic analysis of planar RC structures (Part I). *Transactions of the Architectural Institute of Japan*, No. 240, 51–62.

Toussi S., Yao J.T.P. and Chen W.F. (1984). A damage indicator for reinforced concrete frames, *ACI J.*, **81**, 3, May–June, 260–67.

Wang M.-L. and Shah S.P. (1987). Reinforced concrete hysteresis model based on the damage concept. *Earthq. Engng and Struct. Dynamics*, **15**, 993–1003.

Williams M.S. and Sexsmith R.G. (1995). Seismic damage indices for concrete structures: A state-of-the-art review. *Earthquake Spectra*, **11**, 2, 319–49.

Chapter 5

Broderick B.M., Elnashai A.S. and Izzuddin B.A. (1994). Observations on the effect of numerical dissipation on the nonlinear dynamic response of structural systems. *Engineering Structures*, **16**, 1, 51–61.

CEB (1985). *Model code for seismic design of concrete structures*. Comité Euro-International du Béton, Bulletin d'Information, 165, Paris.

CEB TG III/6 (1994) *Behaviour and analysis of reinforced concrete structures under alternate actions inducing inelastic response — Vol. 2: frame members*. Comité Euro-International du Béton, Bulletin d'Information, 220, Lausanne.

CEN Techn. Comm. 250/SC8 (1995). *Eurocode 8: earthquake resistant design of structures — Part 1: general rules*. ENV 1998-1-1/2/3, CEN, Brussels.

Colangelo F., Giannini R. and Pinto P.E. (1996). Seismic reliability analysis of RC structures with stochastic properties. *Structural Safety*, **18**, 2/3, 151–68.

Dolce M. and Evangelista L. (1990). A critical appraisal of the seismic code provisions for capacity design of RC buildings. *Proc. 9th Europ. Conf. on Earthq. Engng*, Moscow, **10A**, 13–22.

Dolce M. and Evangelista L. (1991). Calibrazione della resistenza dei pilastri per una corretta risposta sismica degli edifici. *Proc. 5th Italian Conf. on Earthq. Engng*, Palermo, 701–10.

Elnashai A.S. and Izzuddin B.A. (1993). Modelling of material nonlinearities in steel structures subjected to transient dynamic loading. *Earthq. Engng and Struct. Dynamics*, **22**, 509–32.

Elnashai A.S. and Papazoglou A. (1997). Procedures and spectra for analysis of RC structures subjected to strong vertical earthquake loads. *J. Earthq. Engng*, **1**, 1, 121–56.

Fardis M.N. (1994). Damage measures and failure criteria for reinforced concrete members. *Proc. 10th Europ. Conf. on Earthq. Engng*. Ed. G. Duma, Balkema, Rotterdam, 1377–82.

Fardis M.N. and Panagiotakos T.B. (1997). Seismic design and response of bare and masonry-infilled concrete buildings. Part I: bare structures. *J. Earthq. Engng*, **1**, 1, 219–56.

Fardis M.N. (1994). *Background information regarding modelling and the analysis*. Prenormative research in support of Eurocode 8 Report.

Izzuddin B.A. and Elnashai A.S. (1989). ADAPTIC, a program for adaptive large displacement elastoplastic dynamic analysis of steel and composite frames. ESEE Report No. 7/89, Imperial College, London.

Kanaan A. and Powell G.H. (1973). *DRAIN-2D: a general purpose computer program for dynamic analysis of inelastic plane structures*. Earthquake Engineering Research Center, University of California, Berkeley, CA, Rep. EERC-73/6 and EERC-73/22, Apr. 1973 (revised Sep. 1973 and Aug. 1975).

Kappos A.J. (1986). Input parameters for inelastic seismic analysis of RC frame structures. *Proc. 8th Europ. Conf. on Earthq. Engng*, Lisbon, Portugal, **3**, 6.1/33–40.

Kappos A.J. (1991). Analytical prediction of the collapse earthquake for RC buildings: suggested methodology. *Earthq. Engng and Struct. Dynamics*, **20**, 2, 167–76.

Kappos A.J. (1996). *DRAIN-2D/90: program for the inelastic dynamic analysis of plane structures subjected to seismic loading — user's manual*. ESEE Report No. 96-6, Imperial College, London.

Kappos A.J. (1997a). Influence of capacity design method on the seismic response of RC columns. *J. Earthq. Engng*, **1**, 2.

Kappos A.J. and Athanassiadou C.J. (1997). Influence of ductility class on seismic performance of RC structures designed to Eurocodes 2 and 8. *Europ. Earthq. Engng*, (at press).

Litton R.W. (1975). *A contribution to the analysis of concrete structures under cyclic loading*. PhD thesis, Civil Engineering Dept., University of California, Berkeley, CA.

Madas P. and Elnashai A.S. (1992). A new passive confinement model for the analysis of concrete structures subjected to cyclic and transient loading. *Earthq. Engng and Struct. Dynamics*, **21**, 409–31.

Martinez-Rueda J.E. and Elnashai A.S. Confined concrete model under cyclic load, accepted for publication in RILEM *Material and Structures*. October 1995.

Otani A. (1974). Inelastic analysis of RC frame structures. *Am. Soc. Civ. Engrs, J. Struct. Div.*, **100**, ST7, 1433–49.

Papazoglou A. and Elnashai A.S. (1996). Analytical and field evidence of the damaging effect of vertical earthquake ground motion. *Earthq. Engng and Struct. Dynamics*, **25**, 1109–37.

Park Y.J. and Ang A.H-S. (1985). Mechanistic seismic damage model of reinforced concrete. *Am. Soc. Civ. Engrs, J. Struct. Engng*, **111**, 722–39.

Park Y.J., Ang A.H-S. and Wen Y.K. (1987). Damage-limiting aseismic design of buildings. *Earthquake Spectra*, **3**, 1.

Paulay T. (1979). *Developments in the design of ductile reinforced concrete frames*. Bulletin of the NZNSEE, **12**. 1, 35–48.

Paulay T. and Priestley M.J.N. (1992). *Seismic design of reinforced concrete and masonry buildings*. J. Wiley & Sons, New York.

Penelis G.G. and Kappos A.J. (1997). *Earthquake-resistant concrete structures*. E & FN SPON (Chapman & Hall), London.

Priestley M.J.N. and Calvi G.M. (1991). Towards a capacity design assessment procedure for reinforced concrete frames. *Earthquake Spectra*, **7**, 3, 413–37.

Salvitti L.M. and Elnashai A.S. (1995). *Evaluation of behaviour factors for frame–wall RC structures designed to Eurocode 8*. ESEE Report No. 5/95, Engineering Seismology and Earthquake Engineering, Imperial College, London.

Standards Association of New Zealand (1982). (a) *Code of practice for the design of concrete structures* (NZS 3101 — Part 1: 1982); (b) *Commentary on code of practice for the design of concrete structures* (NZS 3101 — Part 2: 1982), Wellington.

Takeda T., Sozen M.A. and Nielsen N.N. (1970). RC response to simulated earthquakes. *Am. Soc. Civ. Engrs, J. Struct. Div.*, **96**, ST12, 2557–73.

Chapter 6

ACI (1996). *Performance based design of RC buildings.* Draft Report ACI-368, American Concrete Institute, Detroit, MI.

ATC21 (1988). *Applied Technology Council — rapid visual screening of building for potential seismic hazards.* Redwood City, 189 pp.

ATC22 (1989). *Applied Technology Council — handbook for seismic evaluation of existing buildings (preliminary).* Redwood City, 169 pp.

ATC14 (1982). *Applied Technology Council — evaluating the seismic resistance of existing buildings.* Redwood City, 370 pp.

Beckingsale C.W. (1980)., *Post-elastic behaviour of reinforced concrete beam–column joints.* Department of Civil Engineering, University of Canterbury, Research Report No. 80-20, 398 pp.

Benzoni G., Ohtaki T., Priestley M.J.N. and Seible F. (1996). Seismic performance of reinforced concrete columns under varying axial load. *Proc. 4th Annual Caltrans Seismic Research Workshop.* California Department of Transportation, Division of Structures, Sacramento, CA.

Calvi G.M, and Kingsley G.R. (1995). Displacement based design of multi-degree-of-freedom bridge structures. *Earthq. Engng and Structural Dynamics*, **24**, 1247–66.

Carr A.J. (1995). *Ruaumoko users manual.* University of Canterbury, New Zealand.

EC8/1/1-3 (1994). *Design provisions for earthquake resistance of structures — general and buildings.* ENV 1998-1, CEN, Brussels.

EC8/1/4 (1995). *Design provisions for earthquake resistance of structures — strengthening and repair of buildings.* ENV 1998-1-4, CEN, Brussels.

ERB, New Zealand National Society for Earthquake Engineering (1996). *The assessment and improvement of the structural performance of earthquake risk buildings.* Draft report.

Hakuto S., Park R. and Tanaka H. (1995). *Retrofitting of reinforced concrete moment resisting frames.* Department of Civil Engineering University of Canterbury, Research Report 95-4, Christchurch, New Zealand.

Japanese Building Disaster Prevention Association (1977). *National standard for seismic evaluation of existing reinforced concrete buildings.* Japan, 131 pp.

Kurase Y. (1987). *Recent studies on reinforced concrete beam–column joints in Japan.* PMFSEL Report No. 87-8, University of Texas at Austin, 164 pp.

Mander J.B., Priestley M.J.N. and Park R. (1988). Theoretical stress–strain model for confined concrete. *J. Am. Soc. Civ. Engrs*, **114**, 8, 1827–49.

New Zealand Government (1979). *Local Government Act, Section 624.*

NZS4203 (1992). *Loading code — code of practice for general structural design and design loadings for buildings, Vol. 1 — code of practice.* New Zealand Standards Association, Wellington, New Zealand.

NZS3101 (1995). *Design of Concrete Structures.* Vols 1 and 2, Standards Association of New Zealand, Wellington.

Paulay T. and Priestley M.J.N. (1992). *Seismic design of reinforced concrete and masonry buildings.* John Wiley & Sons, Inc., New York, 714 pp.

Pessiki S.P., Conley C.H., Gergely P. and White R.N. (1990). *Seismic behaviour of lightly reinforced concrete column and beam–column joint details*. NCEER Technical Report No. 90-0014, State University of New York at Buffalo.

Priestley M.J.N. and Calvi G.M. (1991). Towards a capacity design assessment procedure for reinforced concrete frames. *Earthquake spectra*, **7**, 3, 413–37.

Priestley M.J.N. and Seible F. (1994). Seismic assessment of existing bridges. *Proc. 2nd Workshop on the Seismic Design of Bridges*. Queenstown, New Zealand, **2**, 46–70.

Priestley M.J.N. and Xiao R.V. (1994). Seismic shear strength of reinforced concrete columns. *Am. Soc. Civ. Engrs, J. Struct. Engng*, **120**, 8, 2310–29.

Priestley M.J.N., Seible F. and MacRae G. (1995). *The Kobe earthquake of January 17, 1995 — initial impressions from a quick reconnaissance*. University of California, San Diego, Structural Systems Research Project, Report No. SSRP-95/03, 71 pp.

Priestley M.J.N. (1995). Displacement-based seismic assessment of existing reinforced concrete buildings. Keynote address. *Proc. Pacific Conf. Earthq. Engng*, Melbourne, Australia, November, **2**, 225–44.

Priestley M.J.N and MacRae G.A. (1996). Seismic tests of precast beam-to-column joint subassemblages with unbonded tendons. *J. Prestressed Concrete Institute (America)*, **41**, 1, 64–81.

Priestley M.J.N., Kowalsky, M.J., Ranzo G. and Benzoni G. (1996). Preliminary development of direct displacement-based design for multi-degree of freedom systems, *Proc. SEAOC Annual Conf.*, Hawaii.

Priestley M.J.N., Seible F. and Calvi G.M. (1996). *Seismic design and retrofit of bridges*. John Wiley and Sons, Inc., New York, 686 pp.

Priestley M.J.N. (1997). Displacement-based seismic assessment of reinforced concrete buildings. *J. Earthq. Engng*, ICPress, **1**, 1, 157–92.

Shibata A. and Sozen M. (1976). Substitute structure method for seismic design in reinforced concrete. *Am. Soc. Civ. Engrs, J. Struct. Div.*, **102**, 1, 1–18.

UBC (1994). *Uniform Building Code — structural engineering design provisions*. ICBO, Whittier, California.

Wyllie L.A. (Ed.) (1986). The Chile Earthquake of March 3, 1985. *Earthquake Spectra*, **2**, 2.